THE PHYSICAL BASIS
OF
MUSICAL SOUNDS

THE PHYSICAL BASIS
OF
MUSICAL SOUNDS

by
Joseph Morgan, Ph.D.

Professor of Physics, Emeritus
Texas Christian University
Fort Worth, Texas

ROBERT E. KRIEGER PUBLISHING COMPANY
HUNTINGTON, NEW YORK
1980

ΞB

Original edition 1980

Printed and Published by
ROBERT E. KRIEGER PUBLISHING CO., INC.
645 New York Avenue
Huntington, New York 11743

Printed in the United States of America

Library of Congress Cataloging in Publication Data

Morgan, Joseph, 1909-
    The physical basis of musical sounds.

    Includes index.
    1. Music—Acoustics and physics. I. Title.
ML3805.M62   781'.1   78-5508
ISBN 0-88275-656-7

To my grandchildren

Kathy and Joe

Table of Contents

List of Illustrations

Preface

This book is an outgrowth from a one-semester course given for more than a decade for university music students and some interested science students. Since music is a science as well as an art, a knowledge of the principles and laws that govern the production, emission, and transmission of musical sounds serves to contribute to the overall training, depth of appreciation, and musical enjoyment of the music student and the lover of music. As one who has spent many years as a student of music, the knowledge of the physical basis of musical sounds has afforded a degree of appreciation of instrumental and orchestral music that could not have otherwise been acquired. This text, which is an expansion of lecture notes developed throughout the years, is written from both the musician's and scientist's standpoint and attempts to provide the answers to questions from fine arts students, music students, music performers, and others interested in the science of music.

The coverage of material is designed for a one-semester course for freshmen or sophomores and even for upper level students and could also be used in the one-quarter time allotted in schools that function on the quarter system. The emphasis is on the development and explanation in simple language of the relevant physical laws, principles, and musical phenomena. Physical laws and principles are best understood and provide an effective working knowledge when stated quantitatively. Since these necessitate somewhat of a background we have started the text with the first chapter devoted to basic physical concepts that govern the characteristics of vibrations inherent in the production of all musical sounds. In later chapters these quantitative relationships are invoked to develop, describe and explain various phenomena but we have employed, where necessary, only the most elementary mathematics. For example, the development of the laws governing the vibrations of strings and wind instruments is presented at the most elementary quantitative level and the developments of the just diatonic and equal temperament scales and systems are made on an elementary arithmetic level. Mathematical equations have been kept to a minimum and included are only expressions that have basic pertinence, and many of these are usually of use in a laboratory experiment that may be a part of the course.

After the chapter on basic physical concepts there is a presentation of simple harmonic motion in Chapter 2, a development of mechanical waves and their transmission in Chapter 3 followed by the phenomenon of musical sound wave interference in Chapter 4. Chapter 5 deals with vibrating systems followed by a presentation in Chapter 6 of the significant characteristics of sound, viz., pitch, loudness and quality. The study of intervals, scales and temperament is covered in Chapter 7. Chapter 8 deals with musical instruments but as sources of musical sounds and no attempt is made to give a detailed account as to the physical makeup and construction of each instrument since this is provided by publications especially designed for this purpose. Rather, in this chapter we categorize the various families of musical instruments and describe some of the aspects that are significant in affecting the characteristics of the emitted musical sounds. The book ends with the chapter on the acoustics of music halls and auditoriums, the essentials of which a student of music or performer should necessarily always be aware..

Each chapter ends with a set of carefully chosen problems designed to aid in understanding the principles covered and at times to augment them by providing the necessary hint as a guide in their solution. The author's experience has shown that the student acquires a much more lasting understanding of the material covered by working the problems. For example, primarily by applying the arithmetic computations of the musical intervals does one acquire the definitive feeling of a musical interval which is the basic unit of harmony and melody. Answers to the odd-numbered problems are given at the back of the text.

The author wishes to make the following grateful acknowledgments: to many of his former students whose enthusiasm and continual stimulation during class helped the author in evolving this book; to Norma Baptist for typing the manuscript; and to his wife, Edith L. Morgan, for her unfailing encouragement and valuable advice throughout the entire writing of the manuscript.

Joseph Morgan
Annandale, Virginia
April, 1978

CHAPTER I

Fundamental Physical Concepts

All musical sounds have one thing in common; their source is a vibratory unit such as a string, a reed, an air column, a drum head, a loudspeaker diaphragm, a bell, a vocal cord, or any mechanical pulsation. The transmission of these vibrations takes place, as will later become evident, through an elastic medium such as the air. The auditory sensation results when these sound vibrations enter the ear. They then initially set the ear drum (tympanic membrane) into vibration which generates nerve impulses that eventually reach the appropriate brain center, thus giving rise to the physiologic sensation of hearing. We must therefore study the characteristics of vibratory motion if we are to understand the physical basis of musical sounds. This in turn involves our first considering some significant physical concepts that are basic to the study of vibratory motion and this we do in this first chapter.

1.1. Displacement

One of the quantities that are essential in the description of motion is displacement. When a violin string is bowed or plucked the string experiences a displacement and so does the bow or the finger used to pluck the string. When the diaphragm of a loudspeaker is vibrating, to-and-fro displacements are occurring. In general when a body moves from one point to another a displacement has taken place. If the displacement is along a straight line, it is a linear displacement. If the displacement is along a curve, such as the bob of a simple pendulum, we have an angular displacement. In either case, in addition to the magnitude that every displacement possesses there must also be specified a direction or sense of the displacement. Thus we specify a linear displacement as a number of length units north-eastward. For an angular displacement we must tag on the clockwise or counterclockwise sense. Quantities that require for their complete specification both a magnitude and a direction or sense are called *vector* quantities. On the other hand a quantity that does not have a direction or sense associated with it, as time, and is completely specified by giving its magnitude is called a *scalar* quantity. We shall meet quantities of both types as we proceed with the various developments.

The measurement of displacement is made in terms of some unit of length. In the French or metric system this standard of length is the *meter* (m), whose abbreviation is indicated in the parenthesis. The meter is conveniently subdivided into one hundred parts each of which is called a *centimeter* (cm) which is further subdivided into ten parts each of which is called a *millimeter* (mm). To indicate lengths or displacements much larger than a meter it is convenient to use the kilometer (km) which is one thousand meters.

1.2 Time.

Any description of motion necessarily involves the fundamental physical quantity *time*. An interval of time is measurable in terms of any periodic device. Our standard unit of time, the *second*, was originally defined in terms of the rotation of the earth about its axis resulting in the observed sun's daily apparent movement across the sky from east to west. The interval of time between two successive passages of the sun across the earth's meridian at any place is arbitrarily called a solar day. Since this varies somewhat throughout the year the average of all solar days in a year is the more representative value and is called the *mean solar day*. This interval of time is divided in 24 hours (hr) with each hour divided in 60 minutes (min) and each minute into 60 seconds (sec.). The product 24x60x60 equals 86,400 and the second was defined as 1/86,400 of a mean solar day.

However, the rotation of the earth about its axis does not have a constant period of rotation, its speed of rotation decreasing very slowly, and is therefore not an ideal standard. In 1960 the General Conference on Weights and Measures adopted a more refined international definition of the standard and since then atomic clocks have been proposed with an accuracy of about one part in ten billion. However, for all but the most precise purposes, the new definition is equivalent to the old one and we shall consider the second as the 1/86,400 of the mean solar day.

1.3 Speed and velocity.

From the two fundamental quantities displacement and time we can form several derived quantities one of which is speed or velocity which are essential in our description of motion. By *speed* we mean *distance or displacement traversed in a unit of time.* For example when an automobile moves 30 miles or 48.3 kilometers in one hour we express the speed as 30 mi/hr or 48.3 km/hr the derived quantity having a unit which is compounded by a displacement unit and a time unit.

Speed is a scalar quantity and requires only a magnitude and a unit to specify it completely. On the other hand we may also express the motion of the automobile by using the word velocity and then we must also indicate a direction since velocity is a vector. By *velocity* we likewise mean distance or displacement traversed in a unit time but additionally in a specified direction. Hence we reserve the word speed to specify only the magnitude of velocity. For the above automobile we would say the velocity is 30 mi/hr or 48.3 km/hr eastward or southward.

An object that moves with uniform speed covers equal distances in equal intervals of time so that the distance covered per unit of time is the same for each succeeding unit of time. On the other hand a body may slow down at some places and speed up at others and we have what is known as an acceleration which we define in the next section. Nevertheless, if we divide the total distance traversed by the total time required to cover this distance we obtain the average speed or average velocity if the direction is also specified. It is convenient to express this in symbolic form as

$$v_{av} = s/t \qquad (1.1)$$

where v_{av} is average speed or velocity, s is the entire distance traversed and t is the corresponding total time that has elapsed.

It should be clear that if a body were to travel with a uniform speed equal to its average speed, as determined by Equation 1.1, it would cover the same total distance in the same total time. For example, if a body covers 30 miles in the first hour and 40 miles in the second hour the average speed is 35 miles per hour. At a constant speed of 35 miles per hour the body would cover 70 miles in two hours.

1.4. Accelerated Motion

Where the motion involves a change in velocity from point to point we describe this as *accelerated motion.* We then define *acceleration* as *the time rate of change of velocity.* If the motion is along a straight line in a given direction and the body is increasing its velocity from point to point we have a positive acceleration. On the other hand, if the velocity decreases from point to point the acceleration is negative and we may also refer to it as a *deceleration.*

Acceleration is a derived quantity compounded also from the fundamental quantities distance or displacement and time. Since velocity is a vector and time is a scalar acceleration is a vector quantity. Since acceleration possesses both magnitude and direction a body may experience an acceleration in any one of three ways: only its magnitude may change as when it speeds up or slows down in its motion along a straight line; only its direction may change as when it rotates with uniform speed in a circular path; both its magnitude and direction may change as when it moves with variable speed in a circle.

It is again convenient to use symbols to express the acceleration as the time rate of change of velocity. We then state

$$a = \frac{v - v_o}{t} \qquad\qquad (1.2)$$

where a is the acceleration, v_o is some initial velocity at time zero, and v is the velocity at the end of the time interval t so that the right-hand expresses in symbols the time rate of change of velocity. In uniformly accelerated motion, say motion along a straight line, the velocity of the body changes by equal amounts in equal time intervals and the acceleration in Equation 1.2 is a constant. However, if the velocity does not change uniformly, then Equation 1.2 represents the average acceleration.

The unit in which the acceleration is expressed is simply a compound unit obtained by dividing the unit for velocity by the unit for time. As an illustrative example suppose we consider that a train moving eastward passes a point A with a velocity of 30.0 miles per hour and, after accelerating uniformly for 10.0 minutes, passes a point B with a velocity of 75.0 miles per hour. Let us find the acceleration of the train and the distance between A and B. Applying Equation 1.2 we list

$$v_o = 30.0 \text{ mi/hr}, \; v = 75.0 \text{ mi/hr}, \; t = 10.0 \text{ min.}$$

and substituting

$$a = \frac{75.0 \text{ mi/hr} - 30.0 \text{ mi/hr}}{10.0 \text{ min}}$$

$$a = \frac{4.50 \text{ mi/hr}}{\text{min}}$$

The result states that the train increases its velocity by 4.5 mi/hr each minute. The units of course are correct but it is cumbersome to leave the answer with two different time units and more consistent to convert to one time unit, usually the second. At the same time we may wish to express the miles unit in feet (ft). We do this by using any known conversion factors like 5280 ft = 1 mi, 3600 sec = 1 hour, 60 sec = 1 min and proceed as follows:

$$a = \frac{4.50 \ \text{mi/hr}}{\text{min}} = \frac{4.50 \ \cancel{\text{mi}}}{\cancel{\text{hr}} \cdot \cancel{\text{min}}} \ \text{x} \ \frac{5280 \ \text{ft}}{\cancel{\text{mi}}} \ \text{x} \ \frac{\cancel{\text{hr}}}{3600 \ \text{sec}} \ \text{x} \ \frac{\cancel{\text{min}}}{60 \ \text{sec}}$$

Notice that each of the conversion factors used as a multiplier has the numerical value of unity and therefore does not introduce any alteration in the answer. The units manipulate in algebraic fashion and we have divided out similar units that appear in the numerator and denominator. We are now left with the answer

$$a = \frac{4.50 \ \text{x} \ 5280}{3600 \ \text{x} \ 60} \ \frac{\text{ft}}{\text{sec x sec}} = .11 \ \frac{\text{ft}}{\text{sec}^2} = 11 \ \text{x} \ 10^{-2} \ \text{ft/sec}^2$$

where we have simplified the time statement by using the algebraic equivalent that sec x sec = sec^2 and expressed the number in an equivalent but more scientific fashion since ten raised to the negative exponent 2 is the same as $1/100$ or $10^{-2} = 0.01$.

To complete the problem we now employ Equation 1.1 to obtain the distance traversed. Thus the distance s is

$$s = v_{av}t = (\frac{v_o + v}{2})t \tag{1.3}$$

where for the average velocity we have used half the sum of the initial velocity v_o at the beginning of the time interval and the velocity v at the end of the time interval. Remembering that we wish our distance to come out in feet we shall proceed somewhat differently by making our unit conversions before we substitute into the equation. Hence

$$v_o = 30.0 \ \frac{\cancel{\text{mi}}}{\cancel{\text{hr}}} \ \text{x} \ \frac{5280 \ \text{ft}}{\cancel{\text{mi}}} \ \text{x} \ \frac{\cancel{\text{hr}}}{3600 \ \text{sec}} = 44.0 \ \text{ft/sec}$$

$$v = 75.0 \ \frac{\cancel{\text{mi}}}{\cancel{\text{hr}}} \ \text{x} \ \frac{5280 \ \text{ft}}{\cancel{\text{mi}}} \ \text{x} \ \frac{\cancel{\text{hr}}}{3600 \ \text{sec}} = 110 \ \text{ft/sec}$$

$$t = 10.0 \ \text{min} = 600 \ \text{sec}$$

Then,

$$s = \frac{44.0 \ \text{ft/sec} + 110 \ \text{ft/sec}}{2} \ \text{x} \ 600 \ \text{sec} = 462 \ \text{x} \ 10^2 \ \text{ft}$$

where we have again expressed the result using ten raised to the positive exponent 2 since $10^2 = 100$.

The concept of acceleration is significant in vibrating systems that emit sound. The vibratile element of a musical instrument such as a string or a reed is an oscillatory motion that proceeds with a changing or variable acceleration. We shall study this kind of motion in the next chapter.

1.5. Mass and Force

Ordinarily one would write a separate section for mass and a separate one for force. The preference of heading this section as mass and force is dictated by the fact that without the concept of force one cannot say very much about what is meant by mass. Mass is a property that is possessed by everybody and its characteristics become manifest in the presence of a force. We can say that mass refers to the amount and kind of material present but this is not satisfying; it really doesn't tell us anything about its unique dynamical properties. Let us first say something about a force with which we are more familiar and constantly witness in our every day lives when we lift things, push things, pull on things, etc.

A *force* represents a *push* or a *pull* on a body and this idea is ordinarily accompanied by the muscular effort that we must exert in doing things. In fact some forces may be measured by the muscular effort exerted but this is not a very exact method. A more scientific method is to observe the effect produced on a body when a push or pull is excrted upon it. For example, when you try to life a body which is resting on the ground you exert an upward pull on the body. If the pull or force is great enough the body moves upward. However, if the upward pull is not great enough, the force exerted only tends to lift the body or, as we can say, only tends to change the state of motion of the body from one of rest to one of motion. Why is this? The reason is that there is an opposing downward force acting on the body due to the earth's gravitational pull on it. Only if we exert an upward pull that is greater than the downward pull can we impart upward motion to the body. Likewise, when a push is exerted on a box resting on level ground the force exerted has the effect of tending to change the state of motion of the box. If the box does not move, there is a balancing force acting on the box in the opposite direction due to friction between ground and box. If the push is great enough to produce an unbalanced force, the box moves and its state changes from one of rest to one of motion, i.e., there is produced a change in the velocity or an acceleration. A *force* then *is a push or a pull, that is, an action, exerted upon a body and the effect is to tend to change the state of motion of the body on which the force acts.* A force can also change the size or shape of a body. If we pull on a lump of putty we invariably change its shape. Forces may conveniently be compared or measured by means of a spring balance which consists of a coil elastic spring enclosed in a metal case with one end of the spring attached to the case. The other end is attached to a pointer which is free to move so that it traverses calibrated scale markings on the case.

From what we have already stated it should be clear that it takes a force to set in motion a body that is initially at rest, i.e. to accelerate it, or to stop a body initially in motion. From this the inference follows that a body offers resistance to any change in its state of rest or motion. A body then is inert mechanically and possesses a property known as *inertia*. We may rigorously now define *inertia* as *that property of a body that offers resistance to any change in its state of motion or that requires a force to change its state of motion.* This property is measurable and the numberical measure of inertia is what is known as *mass.*

Mass is the third fundamental physical quantity which, along with the other two, length and time, afford a basis for the development of a consistent system of units. In the metric system the magnitude of a mass is obtained by comparing it with some standard mass, the *kilogram* (kg) which is the mass of a cylinder of platinum kept at the International Bureau of Weights and Measures at Sevres, France. This mass unit is conveniently divided into one thousand parts each of which is known as a *gram* (gm).

1.6. Relationship Between Force, Mass and Acceleration.

There is a significant relationship between the force that acts on a mass and the acceleration that the mass consequently experiences. From what we have already said we may infer this relationship but only in a qualitative fashion. Since mass possesses the property of inertia it follows that a force is required to change its state of rest or motion, i.e. to accelerate it. Also, the greater the intertia, the greater is its measure mass and the greater the force required to accelerate a body by the same amount. Also, if the same force is applied successively to a large mass and small mass the smaller mass will experience a greater acceleration. When such experiments are performed under controlled conditions and accurate instruments are used to measure the forces, the masses, and accelerations we obtain the quantitative relationship.

$$F = ma \qquad (1.4)$$

where F is the force, m is the mass and a is the acceleration. Equation (1.4) is the relationship that is basic to all studies of motion and is known as Newton's Second Law of Motion.

It says that when the mass is kept constant the force is directly proportional* to the acceleration so that if the force is doubled the acceleration is doubled, or if the force changes by a given fraction, the acceleration will change by the same fraction. It also says that if the force is kept constant, the mass and acceleration are inversely proportional to one another so that if the mass is increased by a given fraction, the acceleration decreases by the same fraction and vice versa.

Now force is by nature a vector, requiring a stipulation of magnitude as well as direction. We also know that the mass of a body is the same anywhere in the universe for it will accelerate by the same amount under the action of a given force no matter where the experiment is performed.

Equation 1.4 is basically a vector equation and in effect says three things: first it says that the magnitude on one side of the equation must be equal to the magnitude on the other side; second it says the direction of the acceleration must be the same as the direction of the force; and third it says the units used to express the force must be the same as the units used to express ma.

We must now obtain our derived unit of force so that there will be an equality in units. In the metric system if m is expressed in grams and a in centimeters per second we have a derived unit for the force which is a *gm.cm/sec²*. This is called the *dyne*. If m is expressed in kilograms and a in meters per second per second the derived unit for the force is the kg.m./sec² which is called the *newton* (N). One *dyne* is that amount of force that will accelerate a one gram mass one centimeter per second per second. One *newton* is that amount of force that will accelerate a one kilogram mass one meter per second per second. The relation between the newton and the dyne is obtained as follows:

$$1 \text{ newton} = 1 \text{ kg.m/sec}^2 = 1000 \text{ gm x } 100 \text{ cm/sec}^2$$
$$= 100,000 \text{ dynes} = 10^5 \text{ dynes}$$

where we have expressed the result as 10 raised to the fifty power since $100,000 = 10^5$.

*When we say that some quantity x is directly proportional to another quantity y we mean that the ratio of x to y always remains constant. We state this symbolically as x∝y. Therefore x/y = C, or x = Cy where C is the constant ratio.

1.7. Distinction Between Mass and Weight.

We may apply Newton's second law to a mass situated at some place on the surface of the earth. It is then acted on by the gravitational force exerted by the earth on any body. This force which the earth exerts on the mass is called the weight of the body, w, which we can substitute for F in equation 1.4 and obtain

$$w = mg \qquad (1.5)$$

where we have substituted for the symbol a the symbol g to represent the gravitational acceleration that the mass experiences due to the earth's force of attraction for the mass near the earth's surface. Since the earth is not perfectly spherical and also spins around its axis, g varies somewhat over the surface of the earth, depending upon latitude. However this variation from the poles, where the earth is flattened, to the equator is only 0.052 m/sec^2 and g for our purposes may be considered constant and equal to 9.80 m/sec^2 or 32.0 ft/sec^2.

Since the weight of a body represents the gravitational force it is a vector which varies somewhat over the earth's surface. Neglecting this variation, then when the weight of a mass is determined on an equal arm balance by comparison with a known standard weight, Equation 1.5 affords the means by which we may determine the mass of a body by the method of weighing.

1.8. Newton's Laws of Motion

We have already discussed Newton's second law of motion which is the basic quantitative law relating the force, mass and acceleration. We must additionally point out that F in this law is really an unbalanced force. For example if two diametrically equal and opposite forces act on a body at rest, the effect on the state of motion of the body is the same as if no force were acting and the body remains at rest and does not accelerate. Only if one of the forces is greater than the diametrically opposite one is there a tendency to change the state of motion of the body and, in that case, the difference or resultant force is the value that is to be substituted for F in Equation 1.4. Hence it is well to keep in mind that it is the unbalanced or resultant nonzero force that imparts an acceleration a when acting on a mass m.

We would be remiss in our coverage of the basic concepts of motion if we did not at least briefly discuss the first and third laws of Newton.

The first law states that *a body at rest remains at rest, and a body in motion continues to move with a constant velocity in a straight line, unless, in either case, it is acted upon by an unbalanced force.* This law states that it takes an unbalanced force or a nonzero resultant force to change the state of rest of a body. For example, a box resting on a platform is at rest simply because the forces acting on the box are balanced; gravity exerts a downward force on the box but the platform exerts an equal and opposite upward force on the box, so that the resultant vertical force is zero and the box remains at rest. If we apply an upward force greater in magnitude than the downward gravitational force, we would change the state of motion of the box from one of rest to one of upward motion. Again, if we should desire to start the box moving in a horizontal direction, we should have to provide a horizontal force of sufficient magnitude that it would be greater than the opposing maximum frictional force existing between the box and the platform. Then the nonzero resultant force acting horizontally would impart motion to the box in the direction of the unbalanced force. We have already indicated in Section 1.4 that a body that is moving along a curved path is accelerated. Hence if a body is moving along a straight line with uniform speed it will continue to do so unless it is made to increase or decrease its speed along that line or is made to curve; in either case, an unbalanced force acting along the straight line or one having a sidewise thrust would be required. A little thought will bring one to the realization that the first law of Newton is really contained in his second law.

The third law states that *to every action or force there is an equal and opposite reaction or force.* As an example consider a mass supported by a cord above the earth's surface. There is a downward force that is exerted by the earth on the body and the equal and opposite force exerted by the body on the earth. Either of these may be termed the action and the other is termed the reaction. Another pair of action-reaction forces is the downward pull of the mass on the cord and the equal and opposite upward pull of the cord on the mass. It is important to realize that forces always occur in pairs as action and reaction forces but that these never act on the same body. If you exert an action force by pushing your hand against the wall, the wall exerts the equal and opposite reaction force on your hand.

1.9. Elasticity

The property of elasticity possessed by bodies is significant in the generation and transmission of sound. We define *elasticity* as *that property of a body by virtue of which it tends to return to its original configuration, size or shape, when the force which has deformed it is removed.* Some bodies or media are highly elastic such as steel or a coil spring which return to their original configuration after a distorting force is removed. Other materials such as putty or wax are extremely inelastic. It proves to be true that the speed, v, with which sound is transmitted through a medium depends on two physical properties of the medium viz. an elasticity factor E and the density ρ which is the mass per unit volume. This relationship is

$$v = \sqrt{E/\rho} \tag{1.6}$$

For example the speed of sound in air at zero centigrade degrees temperature is about 332 m/sec (1088 ft/sec), while in water it is about four times as much and in steel it is about fifteen times as much.

1.10. Work, Energy and Power

Whenever the action of a force on a body results in the motion of the body then we say that work has been accomplished. The amoung of work done is given by the *product of the force and the displacement of the body in the direction of the force.* Thus if W is the work and s is the displacement then

$$W = F.s \tag{1.7}$$

Work is a scalar quantity and its derived unit is obtained by the product of the units of force and displacement. In the metric system if F is in dynes and s is in centimeters, the unit of work is the dyne.cm which has been named the *erg*. An erg is a very small amount of work; lifting a one gram mass vertically a distance of one centimeter against gravity involves exerting a force of 980 dynes through a distance of one centimeter or an expenditure of 980 ergs of work. The more practical unit usually employed for work is the *joule* which is ten million ergs or 10^7 ergs. If the force is expressed in newtons and the displacement is in meters then the newton-meter is the unit of work. Since newton is 10^5 dynes and a meter is 10^2 cm then the newton-meter is 10^7 ergs or a joule. This illustrates one of the advantages of expressing forces in newtons and displacement in meters.

Energy is defined as *a measure of the capability of doing work.* If a body or an agent is able to do work we say it possesses energy. This capability may be due to one of two physical situations or conditions. In the one situation a body can do work if it is elevated with respect to the surface of the earth because if released it is capable of doing work in falling to the earth's surface; or a stretched or compressed spring is able to do work when released. In each case work had to be performed either to place the body in the elevated position or to stretch or compress the spring, and the equivalent amount of energy has thus been stored in the system. Since it is possible to retrieve this energy it is called *potential energy*. In the case of the elevated body the energy is termed gravitational potential energy; in the case of the spring we have an illustration of elastic potential energy. In each case the body possesses energy because of its position or relative position of its parts. These examples illustrate mechanical potential energy. Similarly there exist other forms of potential energy such as electrical potential energy and magnetic potential energy. In the other situation a body can do work because it is in motion such as an arrow or a bullet in flight. The energy that a body possesses by virtue of its motion is called *kinetic energy*. In both situations the potential energy or kinetic energy that a body or system possesses is measured by the work it can do so that the unit in which energy is expressed is also the erg or the joule.

Often we are concerned with not only the amount of work that an agent can do but also with the speed with which the work is or can be accomplished. If the same amount of work can be performed by two agents, but one of the agents accomplishes the work in one-half the time it takes the other agent, it may be more advantageous to compare them on the basis of the speed with which the work is accomplished. The time rate at which work is done is called *power* and we can set this down symbolically as

$$P = W/t \qquad\qquad (1.8)$$

where P is the power, W is the work and t is the time. The derived unit for power is therefore the erg per second or the joule per second. This latter unit, the joule per second, has been given the more familiar name the *watt*. A 60-watt light bulb, for example, uses 60 joules of energy each second it is consuming the electrical energy to produce light. From Equation 1.8 it is seen that a unit of power multiplied by a unit of time yields a unit of energy so that the watt.sec is the same as the joule of energy. In fact the electric companies charge for the use of energy the cost being a few cents per kilowatt-hour. Since a kilowatt is 1000 watts and an hour is 3600 seconds, a kilowatt-hour is 3.60×10^6 joules of energy. One watt is really a small amount of power and in dealing with sound the power involved is not much more than a watt. The quantities work, energy and power are scalars possessing only magnitude for a complete specification.

Problems

1. 1. Define a vector quantity, define a scalar quantity, and tabularize in two columns the vector and scalar quantities that we
have discussed in this chapter.

1. 2. Express the following displacements in centimeters (cm) and
millimeters (mm) and state your answers also as three-digit
numbers times 10 raised to the proper power or exponent: 1.50
meters, 300 meters, 5.75 kilometers, 500 kilometers, 0.00125
meter. (Hint: The answer for similar problems such as 25 kilometers is 25 km = 25000 m = 2,500,000 cm = 2.50×10^6 cm =
25,000,000 mm = 2.50×10^7 mm)

1. 3. From a point A draw a vector that indicates a displacement of
3.00 cm eastward to a point B. Terminate this line at B with the
head of an arrow pointing eastward. AB now represents a displacement vector of 3.00 cm to the east. From B draw a displacement vector 4.00 cm northward to a point C. Now measure the
displacement vector from A to C and the angle that AC makes
with the east direction. The vector AC is called the resultant
vector displacement of the two component vector displacements AB and BC.

1. 4. A violinist moves the length of his bow, which is 75.0 cm,
across a string in 3.00 seconds. Find the average speed of the bow.

1. 5. Why is the concept of acceleration of importance in the study
of the physical basis of musical sounds?

1. 6. A motorcycle changes its velocity from 30.0 mi/hr northward
to 90.0 mi/hr, in the same direction in 20.0 sec. Find the acceleration in ft/sec^2. Note: 30.0 mi/hr is equal to 44.0 ft/sec.

1. 7. In problem 1.6 find (a) the average speed in ft/sec, and (b) the
distance covered during the period of acceleration.

1. 8. Convert 50.0 km/hr to the proper number of ft/sec. Note: 2.54
cm = 1 inch.

1. 9. Why is the concept of (a) mass, and (b) force of importance in
the study of the physical basis of musical sounds?

1.10. You are traveling at 2.75 km/sec in your automobile and you
apply the brakes which slow you down to 0.75 km/sec in 5
seconds. Find the magnitude of the deceleration.

1.11. In Problem 1.10 find (a) the average speed, and (b) the distance
covered during the period of deceleration.

1.12. If the mass of the automobile and contents in Problem 1.10 is
1500 kg, what constant braking force was applied to the auto-

mobile? Express your answer in newtons and in dynes.

1.13. In Figure 1.1 is shown a mass m suspended by a fine thread A and a similar thread B suspended from the mass. If a slowly acting downward force is applied on B, the thread A will always break. However, if the downward applied force is in the form of a very rapid jerk, the thread B always breaks. You can easily verify these results by doing this experiment. Explain how Newton's three laws are illustrated by this experiment.

1.14. A Musical instrument has a mass of 600 gm and a weight of 5.88 N on the earth. How much is its mass and weight on the surface of the moon where the gravitational acceleration is one-sixth that on the earth?

1.15. How does the physical property of elasticity enter into the study of sound?

1.16. Define the following: erg, joule, watt.sec, kilowatt.hr.

1.17. A constant unbalanced force of 100 N acts horizontally on a mass and moves it horizontally a distance of 5.50 m for 20.0 sec. Find (a) the work performed in joules, and (b) the power expended.

the mass possesses it continues in its upward motion above 0 in accordance with Newton's first law of motion. When the mass is moving upward above 0 the spring is under compression and exerts a downward unbalanced force which increases again in accordance with Hooke's law, bringing the mass to instantaneous rest and P'. At P' the downward unbalanced force is a maximum and the mass is acted upon to move downward. It then increases its speed downward, over-shooting the equilibrium position 0 and the cycle of events repeats, resulting in the oscillatory motion. We note that when the mass is below the equilibrium position 0 the spring force acts upward and when the mass is above 0 the spring force acts downward. In either case the spring force tends to restore the body to its equilibrium position 0 which is why it is named a restoring force in this oscillatory motion.

The motion we have been describing is called *simple harmonic motion* which we may now define as *straight-line to-and-fro motion in which the force is directly proportional to the displacement of the vibrating mass from the equilibrium position and always directed toward that position.* In this kind of motion we see that at the end points of the motion P and P', the restoring force and hence the acceleration of the mass is a maximum and its speed is instantaneously zero while at the equilibrium or midpoint position 0 the restoring force and acceleration are zero and the speed of the mass is a maximum. The motion is periodic, repeating itself in equal intervals of time. The time for a complete vibration or oscillation is known as the *period*, T, which is the time, for example, it takes the mass to oscillate over the path 0 P P' 0. One complete oscillation is known as a cycle so the units used for the period is seconds per cycle. The reciprocal of the period has the units cycles per second and represents the *frequency*, f, of the motion. Thus

$$T = 1/f \quad \text{or} \quad f = 1/T \qquad (2.2)$$

A frequency of one cycle per second is called one *hertz* (abbreviated Hz). The maximum displacement, A, of the mass, above or below the equilibrium position is known as the *amplitude* of the motion.

2.3. Graphical Representation of SHM

Consider Figure 2.2 which contains a circle that is radially divided into eighteen equal sectors the angle between any two adjacent radii being 20 degrees. To the right of the circle is a set of coordinate axes, the horizontal axis or abscissa, passing through the center of the circle, being represented by the angle 0 that each radius makes with the positive horizontal radius as indicated. Projecting horizontally the end point of each radius to the corresponding angle on the 0 axis yields the points through which an important curve called the sine curve is drawn. If we consider that a particle is rotating around the circle with uniform speed we can see that the projections forming the sine curve are the same as the projections on the vertical diameter PP′ of the circle but have been spread out laterally to form the sine curve. If a mass is considered moving along PP′ coincident with the projection or shadow of the rotating particle then there must be a force associated with this motion. This mass would perform a straight line to-and-fro motion, and further investigation, which we shall here not pursue, proves that the mass performs simple harmonic motion. In fact we may alternatively state that SHM is the motion of the projection on a straight line of a point that rotates with uniform circular motion, the straight line being in the plane of the circle.

From the above development we emerge with the significant result that SHM is representable by a sine curve. In Figure 2.2 only one complete cycle of the sine curve is shown corresponding to one complete revolution of the rotating particle or one complete vibration of the SHM. For continuous vibrations the motion is periodic and so is its graphical representation. The vertical values of the sine curve are also the vertical values of the simple harmonic motion and represent the displacement y which, as we have already seen, has the maximum value A called the amplitude (see Figure 2.1). It is helpful and important to represent the sine curve by the expression

$$y = A \sin \theta \qquad\qquad (2.3)$$

where y is the displacement, A is the amplitude and θ is the angle in degrees. For example, when $\theta = 0°$, $\sin 0° = 0$ and y = 0. When $\theta = 20°$, $\sin 20° = 0.342$ and y = 0.342A. When $\theta = 90°$, $\sin 90° = 1$ and y = A. When $\theta = 240°$, $\sin 240° = -0.342$ and y = -0.342A. When $\theta = 270°$, $\sin 270° = -1$ and y = -A. Other values for the sine of various angles are given in Table A.1 in the Appendix. It is often useful to express the angle θ in terms of the time t and the frequency of oscillation f. This we can do by realizing that a completion of one cycle spreads over an angle of 360° or 2π radians where one radian is approximately 57.3°. Then 2π radians/cycle times f cycles/sec times t sec yields 2π ft radians which we may substitute for θ and have as another form of Equation 2.3

$$y = A \sin 2 \pi ft \qquad (2.4)$$

The sine curve can therefore just as well be plotted with t as the horizontal axis.

2.4. Significance of SHM in Musical Sounds.

A simple harmonic motion we note is a periodic motion that is executed with a single frequency. For example vibrations of the prongs of a tuning fork execute almost perfect SHM with a representation like that in Figure 2.2. The same is true with the sound emitted by the whistle formed with the human lips as well as some other sounds. However most sounds produced by the variety of instruments composing a symphony orchestra or the sounds from the human voice box are not single frequency sounds. We shall later see that the sine curve is a representation of a single-frequency sound wave, as well as a representation of single-frequency light waves or electromagnetic waves or mechanical waves. When the wave representation contains a single frequency we call it a *simple wave*. Most sound waves contain more than one frequency with different amplitudes and are called *complex waves*.

Figure 2.3 illustrates the simple and the complex waves. The sine curves, drawn as a function of the time t, contain two cycles of a simple harmonic wave a, a simultaneously occurring wave b containing six cycles or having a frequency three times that of a, and their combined or resultant wave c which is also periodic. Curves a and b represent simple waves; curve c represents a complex wave whose frequency is three times that of a and whose amplitude is two-thirds that of a. The waves a and b, representing two sound disturbances traveling through the same medium at the same time, superpose to form the resultant disturbance c whose graphical form is obtained by algebraically adding the vertical displacements at every value of t. In this particular case we say that the complex wave c is composed of a fundamental wave a and its third harmonic whose amplitude is two-thirds that of the fundamental. The precise shape of the complex wave, as we can discern, depends on the frequencies and relative amplitudes of the components. In fact there is a general theorem which was devised by the mathematician Fourier (1769-1830) and experimentally determined by the physicist Helmholtz (1821-1894) that any complex periodic vibration or wave may be built up or represented by the superposition of a number of suitably chosen simple harmonic vibrations or waves. Since most musical sounds have complex wave representations they may be considered composed of simple harmonic waves. We therefore concentrate on learning the properties of simple harmonic motion and then apply the *Fourier Theorem*, as it is called, to the analysis of complex periodic harmonic motions. We shall later learn that the musical characteristic known as quality, that enables a person to distinguish the sound of one musical instrument such as a clarinet from that of a violin, is explainable chiefly by the differences in the wave forms which in turn depend upon the number and relative amplitudes of the component harmonics.

We can apply the Fourier theorem to the case illustrated in Fugure 2.3 as follows:

Let y_a, y_b and y_c represent the displacements of the three waves. Then

$$y_a = A \sin 2 \pi \text{ ft}$$

$$y_b = (2/3) A \sin 3(2 \pi \text{ ft})$$

$$y_c = y_a + y_b = A \sin 2 \pi \text{ ft} + (2/3) A \sin 3(2 \pi \text{ ft})$$

The sin curves, as representation of waves, are called *sinusoidal* or *harmonic curves*, and the term *harmonic motion* actually stems from the early application of harmonic curves to the analysis of sound vibrations.

2.5 Phase and Phase Difference

An important term that is used in SHM and in combinations of waves is phase. In equation 2.3 the angle θ is known as the *phase* angle or simply the *phase* of the motion. Phase may be thought of as a value that indicates the position and the direction of motion of a vibrating body with reference to some given point in its motion. For example, referring to Figure 2.2, where we have considered the median point (center of circle) as the reference point, the point on the sine curve at $\theta = 60°$ represents the fraction 1/6 period that has elapsed since the body last passed the median point in its upward motion. We have here expressed the phase angle as a fraction of the period of the motion. If a point other than the median point is considered as the specified reference point then the phase is expressed with respect to that reference point. Referring again to Figure 2.2 we often indicate a phase difference between two points such as the phase difference of a quarter of a period between the points at $\theta = 90°$ and $\theta = 180°$. In Figure 2.3 we have drawn the third harmonic so that its starting point is in phase with that of the fundamental curve a. If we were to have drawn curve a so that it was shifted to the right by, for example, an elapsed time equivalent to 45° then b would have differed in phase from a by one-eighth period. In that case the form of the resultant wave form c would have changed. We shall be indicating the effects of phase in a later chapter.

Problems

2. 1.   In your own words, and without copying verbatum the phraseology used in the chapter, write a definition of the following: Simple harmonic motion, amplitude, Hooke's law, cycle, frequency, period, harmonic curve, simple wave, complex wave, superposition, Fourier theorem.

2. 2.   List the significant properties of simple harmonic motion.

2. 3.   An elastic spiral spring is supported vertically and the stretch of the spring is observed for a series of suspended masses as follows: for 60 gm, 1.2cm; for 80 gm, 1.6cm; for 120gm, 2.4cm; for 150 gm, 3.0cm. On graph paper plot the force (weight of each mass) F in dynes as ordinate and the elongation or stretch d as abcissa. Draw the graph that passes through the points and find the value of the spring constant. Use the value that the weight of 1 gm is 980 dynes. (Hing: K is obtained by computing the slope $\Delta F/\Delta d$ where ΔF stands for any change in F and Δd is the corresponding change in d.)

2. 4.   A tuning fork is vibrating with a frequency of 512 Hz. What is its period?

2. 5.   Sound vibrations are detectable by the normal human ear from about 20 Hz to about 18000 Hz. What are the corresponding periods of the motions?

2. 6.   The circle in Figure 2.2 is known as the *circle of reference*. What reasons can you give for this designation?

2. 7.   On a sheet of graph paper draw a sinusoidal curve that repeats for two periods. On the same sheet, using the same axes, draw another sinusoidal curve whose starting point is in phase with the first curve but whose period and amplitude are each one-half those of the first curve. By the method of superposition graphically find the resultant or complex curve. State the equation that gives the displacement as a function of the amplitude A, the frequency f and the time t.

2. 8.   Repeat Problem 2.7 but with the second sinusoidal curve
        having its first zero value 45° out of phase with the starting
        point of the first curve.

2. 9.   What is the significance of the principle of superposition, as
        exemplified by problems 2.7 and 2.8, in the study of musical
        instruments?

2.10.   What does the word *simple* signify in the designation simple
        harmonic motion?

CHAPTER 3

Mechanical Waves
and Their Transmission

3.1. Mechanical Waves

Mechanical waves are waves in deformable or elastic media such as sound waves, water waves, and waves on a string. The wave motion is one of the two principal means by which energy may be transmitted from point to point and does not involve a physical transfer of matter from one point to another. The other means of transferring energy from one point to another is by a motion of particles or matter such as the transfer of the kinetic energy of the particles of the wind to the rotation of a windmill. In addition to mechanical waves there are also electromagnetic waves, such as light and radio waves, with which we shall not be concerned in our treatment in this book. Although all waves have significant properties in common we shall be primarily concerned with mechanical waves and in particular with sound waves.

We have seen that sound waves require for their production a moving or vibrating source and an elastic medium. The vibrating source causes a displacement or disturbance of the particles of the elastic medium with which the vibrating source is in contact and this disturbance is transmitted successively to the neighboring particles, resulting in the propagation of the wave motion. When the string of a musical instrument is actuated to vibrate, it oscillates in a direction transverse to its length setting up a wave in the medium of the string. Here the particles of the string vibrate perpendicular to the direction of propagation of the wave, which is along the string, and is therefore called a *transverse wave*. The sidewise vibrations of the string are communicated to the air particles which set up a sound wave in which the particles of air vibrate parallel to the direction of wave travel resulting in what is known as a *longitudinal wave*. Both types of waves are mechanical waves and we must understand the mechanism of the production of each in our study of musical instruments.

Let us first consider how a transverse mechanical wave is produced. Such a wave appears when one end of an elastic spring or rope under tension is vibrated in a direction at right angles to the length of the spring or rope. In Figure 3.1a are shown a succession of equally spaced particles of a rope or other medium numbered 0, 1, 2, 3, 4, . . . , and it is to be imagined that there is an elastic spring between the particles to simulate the elastic force that exists between them. The double-pointed arrow indicates that the particle at 0 is subjected to a vertical to-and-fro simple harmonic motion of amplitude A supplied by a mechanical oscillator such as the prong of a tuning fork. When particle 0 starts downward, its motion is transmitted successively and progressively later in time to the neighboring particles. In (b), where 0 has attained its maximum negative displacement, the disturbance has reached particle 4 after a quarter-period, with particles 1, 2, and 3 displaced as shown. After the second quarter-period, particle 0 is at its equilibrium position, particle 4 has its maximum negative displacement, and the disturbance has reached particle 8. In (d), after the third quarter period, particle 0 has attained its maximum upward displacement, particle 4 is at its equilibrium position, particle 8 has reached its maximum negative displacement, and the disturbance has advanced to particle 12. In (e) particle 0 has returned to its starting position, one period of vibration has been completed, and the disturbance has reached particle 16, which is the first particle in step or in phase with 0. Since the vibrating source is simple harmonic, the wave form shown in (e) is a sine curve. It is to be noted that each particle executes the same simple harmonic motion as the source particle 0, there being present a constant progressive phase difference between neighboring particles. This simultaneous vibratory motion of all of the particles results in a transverse wave disturbance that propagates through the medium with a definite speed. It is important to realize that no part of the medium moves along in the direction of the wave motion but only a disturbance or configuration is propagated through the medium. We have already mentioned, as another example, that when water waves proceed outward from their source of origin, floating objects such as cork simply bob up and down and back and forth, but do not move in the direction of the ripples. Of course the shape of the wave disturbance in Figure 3.1 is determined by the manner in which the vibrating source moves. If the source motion were of such a nature as to deliver only a sharp blow to the rope, only a single wave pulse would travel along the rope. Solids are suitable media for the transmission of transverse mechanical elastic waves while gases are not because the elastic forces between gas molecules are too small. In any case, however, it must be remembered that the transmission of a mechanical wave motion is a consequence of the fact that the medium possesses elasticity and inertia. In Figure 3.1e particles 0 and 16 are in phase and we define the distance from one particle and the very next one which is in phase with it as the wavelength λ as indicated in the diagram.

Let us now turn our attention to the mechanism by which a sound wave is produced and its manner of representation. We again refer to the succession of particles, like those in Figure 3.1a, to show the characteristics of a longitudinal wave but let us imagine that the source particle 0 is subjected to a horizontal to-and-fro simple harmonic motion. When 0 moves to the right the particles to the right of 0 are being crowded together, or compressed. This motion is handed on from particle to particle but progressively out of phase. There is thus generated a wave of compression or *condensation* which moves with the speed of sound in the medium. When 0 moves to the left it creates a gap or leaves a vacuum, if the medium is air, and the particles move to fill the gap thus producing a *rarefaction* which propagates to the right with the same speed as the condensation. During one complete vibration or cycle of the motion of 0 there is generated one complete wave. Hence when 0 oscillates back and forth horizontally there is propagated a continuous train of compressions and rarefactions which give rise to the disturbance or configuration called a *longitudinal*, or a *compressional* wave.

The disposition of the particles in the longitudinal wave are shown in Figure 3.2a at the instant when two periods or two wave lengths have been completed. The series of the 16 equidistant undisturbed particles in Figure 3.1a have been expanded to twice that many, 32, in Figure 3.2a. At the instant shown in Figure 3.2a the particles are at the displaced positions shown but it must be remembered that as the source particle 0 vibrates parallel to the direction of wave travel each of the 32 particles performs this same horizontal oscillatory motion. We note that the longitudinal wave is characterized by a series of alternating condensations, indicated as C, and rarefactions, indicated as R. In the regions of compression or condensation the particles are closer together than is normal and are on their way in the direction of wave propagation, while in the regions of rarefaction the particles are farther apart than is normal and are on their way in the direction opposite to that of wave propagation. It is important to notice in Figure 3.2 that the distance between two successive condensations or two successive rarefactions is a wavelength and that the distance between a condensation and its nearest rarefaction is one-half wavelength.

The wave form of a longitudinal wave is not apparent but we can reveal it by laying off the horizontal displacements of the particles in Figure 3.2a from their normal positions in (c) perpendicular to the direction in which they actually occur. Thus the displacements of particles 1 to 8 in (a) from their corresponding normal position in (c) have been plotted vertically upward at their normal position in (b). Displacements of the particles 9 to 16 have been similarly plotted vertically downward. We thus obtain as a representation of a longitudinal wave, the sine curve, which, on recollection, is what we would expect since the vibrating source is of the simple harmonic type. Thus the sine curve may be used as a representation of both transverse and longitudinal waves, a fact that will be helpful in descriptions and discussions.

Since the ear is an organism that is sensitive to pressure changes it is necessary that we relate the form of the pressure changes that result when a longitudinal sound wave proceeds through the air. The graphs in Figure 3.2 are plots of particle displacements or particle movements as a function of distance or time and these movements of the particles give rise to pressure changes or simple sound pressures whose magnitudes depend upon the velocities of the particles. We may also plot the pressure in the waves on the graph of Figure 3.2b but we will not do so to avoid the possible confusion of another curve. Instead we shall describe the curve that represents the variation of pressure and ask the student to draw it in relation to Figure 3.2 as an exercise in the chapter problems. Since the sound pressure in a longitudinal wave in air varies with the air particle velocities, it follows that the sound pressure is also a periodic function of the time or distance. Referring to the displacement sine curve in Figure 3.2b the pressure curve is also a sine curve but one quarter of a wave length out of phase with the displacement curve. Where the particles are closer together at a condensation, the pressure is higher than the average or ambient pressure and has its positive maximum at the point marked C where the displacement is zero. Where the particles are farther apart, at a rarefaction, the pressure is lower than the average or ambient pressure and has its negative maximum at the point marked R where the displacement is again zero.

3.2. Relation between Velocity, Frequency, and Wavelength

We have indicated that the distance between any point in a wave to the very next corresponding point with which it is in phase is the wavelength λ. We have also indicated that a wave motion progresses or propagates with a definite velocity or speed v. We may therefore express v in terms of the wavelength and the period of any periodic wave. Referring to Figure 3.1e, in the time that 0 (and each of the other particles) executes one complete vibration or period T the wave disturbance has moved a distance λ. Therefore if we divide the distance λ by the corresponding time T, we will obtain the expression for v or

$$v = \lambda/T \tag{3.1}$$

or, since by Equation 2.2 $f = 1/T$ we may alternatively write

$$v = f\lambda \tag{3.2}$$

which is the relationship that is required and most often employed. If f is in hertz than v is in centimeters per second, meters per second or feet per second depending on whether λ is respectively in centimeters, meters, or feet. This expression holds for any kind of wave and connects in very simple fashion three of the most significant characteristics of a wave; its speed or velocity of propagation, its frequency, and its wavelength.

The speed of sound in air at 0° C is about 332 meters/sec or about 1088 feet/sec and the frequency range which the normal human ear can detect is about 20 hertz to about 18000 hertz. Using these values in Equation 3.2 yields for audible sound waves the wavelengths from about 0.7 inches or 1.8 cm to about 54 feet or 16 m. On the other hand, wavelengths to which human eyes are sensitive are much smaller and range from about 3.8×10^{-5} cm. to about 7.2×10^{-5} cm.

3.3. Speed of a Sound in Elastic Media

We have stated in Section 1.9, Equation 1.6, that the speed of sound in an elastic medium depends upon the ratio of an elasticity factor to an inertial factor. In a solid the elasticity factor is known as Young's modulus of elasticity which is the constant of proportionality when Hooke's law, in a more general form, is applied to the material, and the inertial factor is the density of the solid. In a liquid the elasticity factor is known as the bulk modulus of elasticity which is similarly defined in relation to the fractional change in volume of the liquid as the longitudinal wave passes through the liquid, and the inertial factor is again the density of the liquid. In a gas the elasticity factor is the pressure of the gas which is defined as the force per unit area existing anywhere in the gas, and the inertial factor is the density of the gas. Since our study, to the greatest extent, involves the speed of sound from musical instruments in air it is useful to write the governing expression

$$v = \sqrt{\gamma\, p/\rho} \qquad\qquad (3.3)$$

where γ is a constant equal to 1.41 for air. If the pressure p is in dynes per square centimeter and the density ρ is in grams per cubic centimeter than v comes out in centimeters per second.

Since a change in temperature of the air (or other medium) changes both the elasticity and inertial factors, the speed of sound in air varies with the temperature, increasing as the temperature increases. At 0° C in air the speed is 332 m/sec (or 1088 ft/sec) and increases by 0.61 m/sec for each centigrade degree increase in temperature (or by 1.15 ft/sec for each Fahrenheit degree increase in temperature). For solids and liquids the change in speed with temperature is small and, in general, negligible.

Table 3.1 lists the values for the speed of sound in some media.

Table 3.1

Speed of Sound in Various Substances

Medium	Temp. (°C)	Speed (m/sec)	Speed (ft/sec)
Air	0	332	1088
Air	20	344	1127
Oxygen	0	316	1036
Helium	0	965	3164
Hydrogen	0	1270	4164
Acetone	25	1174	3849
Benzene	25	1295	4246
Water	15	1437	4711
Silver	—	2680	8787
Steel	—	5000	16400
Glass (pyrex)	—	5170	16950

Several facts are of interest in examining the table. First is to be noticed that generally the speed is less in gases than in liquids and less in liquids than in solids although there are exceptions as exemplified by the gas hydrogen and the liquid acetone. The determining factor of course is neither the elasticity factor by itself nor the inertial factor by itself by their ratio. In comparing air with hydrogen, the density of air is nearly sixteen times that of hydrogen so the speed of sound in hydrogen is nearly four times that in air. In steel the ratio of the elasticity factor to the inertial factor is about 225 times that for air resulting in the speed of sound in steel about 15 times that in air. However, in any given medium the speed of sound is the same regardless of the emitting musical instrument; if this were not so it would be orchestrally catastrophic!

3.4. Properties of Sound Waves.

All waves, transverse or longitudinal, carry and transmit energy. They are also capable of exhibiting the phenomena of *reflection, refraction, diffraction,* and *interference.* In addition, transvers waves are capable of demonstrating the phenomenon of *polarization* which involves the limiting or quenching of some of the transverse vibrations. However, longitudinal waves do not exhibit the phenomenon of polarization and we shall not be concerned with it in sound waves. In this section we discuss reflection, refraction and diffraction. Interference we take up in the next chapter.

When a dimensionally small source capable of radiating sound equally well in all directions emits a sound the disturbance moves outward from such a point source in all directions in space and the energy spreads out spherically from the source. At any time after the sound is emitted we have a spherical surface all points of which are vibrating in phase, and we call such a surface a *wave front.* The energy carried by the expanding wave fronts falls off or decreases inversely as the square of the distance from the source. At a long distance from the source the radius is large, the curvature of the wave front is small and a small section of it may be considered a plane wave front. We may explore the properties of sound waves by using wave fronts or by using the perpendicular to the wave front known as a *ray* of sound. Sometimes it is more convenient to use the wave front and at other times the ray in physical descriptions.

Reflection and Refraction of Sound

Whenever a sound wave, traveling in one medium, falls on a second medium in which the ratio of elastic to inertial property is different from the one in which it is traveling, some of the energy reflects into the first medium and some of the energy enters the second medium. If the sound strikes the reflecting surface perpendicularly the reflected sound will return to its source as a *simple echo*. If the wave, represented by the ray ab in Figure 3.3, is obliquely incident on the interface separating the two media, the direction of the reflected ray bc makes the same angle i, with the normal N to the interface, as the incident ray. The energy that enters the second medium proceeds in the direction bd which is deviated or refracted from the incident direction. Some of this energy is absorbed as it traverses the medium and the remainder emerges in the transmitted direction de which is parallel to the direction ab if the incident and emergent faces of the medium are parallel as drawn in the figure.

Let us first discuss the phenomenon of reflection. In concert halls and auditoriums some sound reaches the audience by reflection from walls, floor, ceiling, pillars, etc. Reflections that take place from differently located surfaces may give rise to multiple echoes. Certain types of materials reflect low frequencies better than high ones and some substances reflect frequencies that are harmonics only because the substance acts like a filter for the fundamental frequency. If the auditorium is equipped with regularly spaced palings, echoes or reflections of sharp staccato sounds arrive at the ear progressively later and at regular intervals thus giving the sensation of the presence of an extraneous frequency or pitch. If the auditorium or hall is constructed with concave highly reflecting surfaces that have geometric focusing properties, the sound instead of spreading in all directions is focused as illustrated in Figure 3.4. In (a) the reflecting boundaries form an ellipse whose geometric property is such that the sound from a Source S at one point of the ellipse is focused at the opposite equivalent point F for all sound rays. In (b) the two concave coaxial surfaces are parabolas whose geometric property directs all sound rays emanating from the source S are reflected parallel to the axis AB and brought to a focus at F. Such surfaces, as well as others (see Problem 3.14) account for the phenomenon of "whispering galleries." However, these effects are more pronounced if the dimensions of the reflectors are much larger than the wavelength of the sound. To avoid such acoustically disturbing effects concavely curved walls or domed ceilings should be avoided in the design of auditoriums or halls. On the other hand a reflecting backdrop of large radius of curvature concave toward the audience is often desirable in projecting the sound where it is wanted.

Where disturbing curved surfaces are already present in a structure the undesirable effects may be minimized by sectioning or breaking up the reflections or by covering the surfaces with absorbing materials. All concavè surfaces tend to concentrate sound energy by reflection but convex surfaces, such as cylindrical pillars, tend to diffuse it. From the above it is clear that reflection of sound is of extreme importance in auditorium and concert hall acoustics. In the study of architectural acoustics it is often desirable to reduce reflection from walls, ceilings, pillars, etc. to avoid the undesired multiple reflections that cause annoying reverberant music reception and unintelligent speech reception. To accomplish this the surfaces may be covered with absorbing materials. We treat this phase of our study in Chapter 9 entitled *Acoustics of Halls and Auditoriums.*

Let us now consider the phenomenon of refraction. In Figure 3.3 the incident ray is entering a medium in which the speed of sound is less and in such a case the angle that the refracted ray, r, makes with the perpendicular to the surface is less than the angle of incidence i. Such would be the case of sound originating in water and emerging in air. In the case where the sound travels from the lower-speed medium into the higher-speed medium, as from air into water, the refracted ray bends away from the normal, on the other side of the undeviated incident ray, so as to result in the refracted ray, now in the water, making an angle with the normal greater than the incident angle. Thus if the density factor or elasticity factor of a medium is not the same in all directions, the direction of sound propagation will change or the wave front, to which the ray is perpendicular, will tilt. As an example, when the ground is covered with snow or over a frozen lake there is a temperature gradient upward with the temperature of the air remote from the ground or lake warmer and hence less dense than the air closer to the ground or lake. Hence the speed of the sound aloft is greater and the wave fronts are refracted downward as shown in Figure 3.5a. Most of the sound energy is confined to the layer of air close to the ground surface or, in the case of the lake, is reflected and skips along as if between two reflecting layers. Hence, if the air is relatively quiet, sounds can be heard at great distances. In warm weather, the temperature gradient is such that the temperature of the air decreases with height above the ground and the speed of sound decreases with height resulting in refraction that tilts the wave front upward as shown in (b). In this case sounds cannot be heard at large distances. For the same reason sounds are better heard at long distances on clear cool evenings. Wind gradients also have effects that are similar. When the wind is blowing in the same direction as the sound propagation direction, the wave front is bent downward and the sound can be heard at greater distances. The opposite results when the wind is blowing in a direction that is opposite to that of the sound propagation. The temperature gradient between the floor and ceiling of an auditorium may give rise to objectionable refraction effects and plays a part in the overall consideration of architectural acoustics. In general, however, refraction effects are of limited importance in musical transmission indoors but may be of greater importance in the rendition of music by artists or orchestras to large audiences out-doors.

Diffraction

The phenomenon of *diffraction* has to do with the bending of a wave disturbance into the region behind an obstacle when a portion of the wave front is intercepted by the obstacle or the spreading of the wave disturbance after the wave has passed through an opening. The effect is illustrated by Figure 3.6. In (a) a plane wave front W advancing to the right is intercepted by an obstacle O. The region between A and B behind the obstacle is the region of the geometric shadow where one might expect to find no wave energy. However experiment reveals that some of the wave energy does appear in the geometrical shadow as if the wave front bends around the obstacle to reach into the region AB. Actually points of the wave front below and above the obstacle act as secondary sources of disturbances which emit secondary wavelet disturbances in all directions and these combine to produce the effect inside the geometrical shadow. A similar situation occurs in (b) where a portion of the wave front W is proceeding through the hole or aperture CD. Here a wave disturbance reaches into the geometrical shadow regions above E and below F by diffraction or apparent bending of the wave fronts. This phenomenon is exhibited by all waves and the amount of diffraction or the extent of this apparent bending or spreading of the waves depends on the size of the wavelength in relation to the obstacle or aperture. If the wavelength is large compared to the dimensions of the obstacle or aperture the wave disturbance effect is large. On the other hand, if the wavelength is small compared to the dimensions of the diffracting elements the amoung of diffraction is small.

A familiar example is the one where a person sits in an auditorium behind a supporting pillar and can hear the sound waves originating by a performer from the stage but is not able to see the sound source or the performer unless he looks around the pillar. The reason for this is that the audible sound waves have wavelengths that are of the order of feet so that there is considerable diffraction into the geometrical shadow of the pillar while the visible light waves have wavelengths of the order of 10^{-5} cm which result in very little diffraction and the casting of sharp shadows. As another example, one can hear what is being spoken in a room if one stands outside the room at one side of an open doorway. One then hears the low frequency or long wavelength sounds best and as the individual approaches the doorway opening he can hear more and more of the higher frequency or shorter wavelength sounds which do not reach as far into the geometrical shadow as the low frequency sounds. The same diffraction effect occurs when musical sounds come from a loud speaker. The low-frequency or long wavelength sounds can be heard on the side lines far off the axis of the speaker as well as on the axis but the high-frequency or shorter wavelengths can be heard only when the listener is on or near the axis of the speaker. To improve this condition two loudspeakers are often employed; a low frequency or long wavelength speaker called a *woofer* and a high frequency or short wavelength speaker called a *tweeter*. The tweeter is ordinarily composed of several speakers disposed in the shape of a fan so as to spread out the short wavelengths which may then be heard over a very wide angle.

Problems

3. 1.   Describe what it is that moves in a mechanical wave and indicate the motional characteristics of the source of the wave.
3. 2.   The transverse wave on a string has a wavelength of 5.00 cm and a speed of 175 cm/sec. With what frequency is the source of the transverse disturbance vibrating?
3. 3.   Find the speed of a sound wave in a glass rod whose Young's modulus is 62.0×10^9 N/m^2 and whose density is 2.50 gm/cm^3. (Hint: the speed is given by $\sqrt{Y/\rho}$ where Y is Young's modulus and ρ is the density.)
3. 4.   Using the information given in the chapter, roughly sketch in the sound pressure wave on the same axis of the displacement wave shown in Figure 3.2b and note where the positions of maximum and minimum pressures occur in relation to the condensations and rarefactions.
3. 5.   Verify the value given in Table 3.1 for the speed of sound in air at 0° C by computing it using Equation 3.3. For p use the

atmospheric pressure 1.013 x 10^6 dynes/cm^2 and for ρ use the density 1.293 x 10^{-3} gm/cm^3.

3. 6. Verify the value given in Table 3.1 for the speed of sound in air at 20° C using the value of 332 m/sec as the speed at 0°C and the appropriate value for the increase in speed with temperature increase given in the chapter.

3. 7. Show dimensionally or by substituting the units that the units for v in Equation 3.3 are cm/sec if p is in dynes/cm^2 and ρ is in gm/cm^3.

3. 8. Taking the speed of sound in air as 1088 ft/sec, what is the wavelength of a sound whose frequency is the standard pitch 440 Hz?

3. 9. The wavelength of a sound wave in a metal rod is 10.0 m. If the speed of the wave decreases from 5000m/sec to 344m/sec as it emerges in the air, what is the wavelength in the air? (Hint: the frequency remains the same.)

3.10. A metal pipe 150 feet long is struck a blow at one end, and an observer at the other end records 0.126 sec as the time interval between the arrival of the wave disturbance through the pipe and that through the air. Find the speed of sound in the pipe, taking the speed of sound in air as 1100 ft/sec.

3.11. From the same point two waves are emitted in phase with frequencies of 12.0 cycles/sec and 15.0 cycles/sec. What is their phase difference at the end of 1.25 sec?

3.12. Two waves travel through the same medium with a speed of 300 m/sec. The frequencies of the waves are 200 Hz and 250 Hz. Find the phase difference at a point 1.50 m distant from a point where the waves are in phase.

3.13. The echo of a sound in air is heard 0.25 sec after it is made. What is the distance of the reflecting surface from the sound source? Take the speed of sound as 344 m/sec.

3.14. The interior of an auditorium is circular in shape and a sound source is located at one place close to the highly reflecting circular wall. Explain the effect of repeated reflections from the wall and the consequent concentrations of sound energy.

3.15. When water waves are interrupted by obstacles or apertures of ordinary sizes diffraction effects may be easily observed. Give an explanation of this.

CHAPTER 4

Interference

4.1. Basic Meaning of Interference

We have learned that a sound wave is a disturbance or configuration that is propagated through an elastic medium and that the sine curve may be used to represent the simple harmonic wave. We have also seen that this sine curve can be plotted to represent the displacement of the particles of the medium or the pressure as a function of the time or the propagation distance. Since a wave motion is a traveling disturbance initiated by a vibrating source whose motion is transmitted from particle to particle it follows that when two or more vibrating sources exist simultaneously in the same medium, the wave produced by each source travels through the medium and produces its effect as it would if the other waves were not present. In other words, each wave proceeds through the medium and produces its effect independently of the presence of the other waves. Therefore the displacement of a given particle of the medium is a resultant displacement or a sum of the displacements produced by the individual waves. The sum of their individual disturbances is an illustration of the *superposition principle*. Thus, if one of the waves traveling alone through a medium were to produce a particle displacement y_1 at some point, and another wave traveling alone through the same medium were to produce a particle displacement y_2 at the same point, the displacement due to the simultaneous action of both waves is, in accordance with the superposition principle, the resultant $y_1 + y_2$. Since the individual displacements y_1 and y_2 may be positive, negative or zero the sum may be larger than either, smaller than either, equal to either, or even zero. This combination of waves, which may result in reinforcement or in partial or complete cancellation, is what is meant by *interference*.

Illustrations of three types of possible combinations of particle displacements giving rise to interference are shown in Figure 4.1. In (a) are shown two waves, 1 and 2, having the same frequency, in phase, differing in amplitude, and moving in the same direction through the same medium. In accordance with the principle of superposition the combined effect of both waves is obtained by adding the ordinates of each at every instant of time t to obtain the resultant wave R. Here the waves reenforce one another at all times and we have a case of *constructive interference. In (b) the two waves are shown progressing 180°,* or half a wavelength, out of phase at all instants of time producing a resultant R that has zero displacement at every instant. Here the waves destroy each other and we have an illustration of *complete destructive interference.* In (c) the waves 1 and 2 are unequal in amplitude and 180° out of phase producing a resultant different from zero and we have an illustration of *partial destructive interference.* When the waves are neither in phase nor 180° out of phase they combine to strengthen each other at some points and neutralize each other at other points. It is important to bear in mind that when interference occurs energy gets to be redistributed and is never destroyed.

The tuning fork affords a simple illustration of the phenomenon of sound interference. In its normal mode of vibration the prongs oscillate in opposite directions as shown in Figure 4.2. When the prongs approach one another, compressions are produced between them and rarefactions are produced above the upper and below the lower prong. A half period later rarefactions appear between the prongs and condensations appear above the upper and below the lower prong. These two series of waves interfere. If the fork, held by the stem in the fingers, is slowly rotated near one ear about the stem axis the wave trains coming from the two prongs interfere destructively at four positions of the tuning fork during one rotation of the fork. The nodal points N in the figure are the stationary places about which the prongs vibrate. Their significance is brought out in our discussion of the vibration of the free end of a bar or rod whose other end is fixed (see Section 5.4).

4.2. Beats.

A kind of interference, which is very familiar and strikingly noticeable in sound, results when two similar waves whose frequencies differ slightly, travel in the same direction through the same medium at the same time. This combination gives rise to a pulsating or throbbing kind of variation in intensity called *beats*. Figure 4.3 illustrates graphically the formation of beats due to the interference of two sound waves whose frequencies are f_1 and f_2 as shown. In (a) it is to be noticed that as time goes on the two waves gradually fall more out of phase until they are a half period out of phase and then they gradually differ in phase less and less until they are again in phase. Adding the waves, in accordance with the principle of superposition, results in the combined wave in (b) which shows that the waves alternately reinforce and partially or completely cancel each other, resulting in audible beats. The time interval between two beats is the period of the beats or the time it takes one of the sound waves to gain a complete vibration on the other wave. Hence the rate of the beats is the difference in the frequency between the two sound waves or

$$\text{Number of Beats/sec} = f_2 - f_1 \qquad (4.1)$$

As an example, if the two frequencies are f_2 = 406 cycles/sec and f_1 = 404 cycles/sec the two vibrations will be in phase twice each second or the number of beats per second is 406 - 404 = 2.

Beats are used to adjust two tones to a common frequency. The closer the generating tones approach each other the smaller are the number of beats per second so that two tones can be tuned to unison very precisely by making the adjustment so there are no beats. Usually a sonorous emitter under test is adjusted until no beats are produced when sounded together with a source of standard frequency. The principle of beats is used in tuning the piano and the organ.

In the production of organ tones certain organ stops, such as the voix celeste, consist of two ranks of pipes, one tuned at a slightly different frequency than the other. When such a stop is drawn then every key that is depressed actuates the emission of tones from two, or more, pipes tuned slightly differently. This results in a slow beating effect or waviness having the tonal color of a vibrato.

As another example of the use of beats in determining the frequency of an unknown audible sound source consider the following problem: When a tuning fork of unknown frequency is sounded together with one whose frequency is 340 cycles/sec there are produced 8 audible beats/sec, and when sounded with another whose frequency is 326 cycles/sec there are produced 6 audible beats/sec. What is the unknown frequency? In the first case the unknown frequency is either 348 cycles/sec or 332 cycles/sec. In the second case the unknown frequency is either 332 cycles/sec or 320 cycles/sec. Therefore the unknown frequency must be 332 cycles/sec.

4.3. Stationary or Standing Waves.

In this section we discuss a type of interference which not only has basic application to sound and musical instruments but is also basic in its application to all kinds of wave motion. It occurs when two similar waves of the same frequency and amplitude travel through the same medium at the same time but in opposite directions. This kind of wave combination results in an interference pattern of *stationary,* or *standing* waves which exhibit regions (points in one dimension, lines in two dimensions or surfaces in three dimensions) where some characteristic of the wave, such as the displacement or pressure in a longitudinal sound wave, is zero, and other regions where this characteristic is a maximum. The regions of zero value are called *nodes* and those of maximum value are called *antinodes* or *loops.* Such a standing wave pattern of displacement nodes and loops are readily made visible when one end of a cord is fastened to a rigid support and the free end is vibrated transversely. The outgoing waves and the reflected waves at the support constitute the oppositely directed waves thus forming the standing wave pattern. The formation of such a system of standing waves can be understood from the series of graphs in Figure 4.4. In (a) are shown two harmonic waves (solid curves) of equal frequency and of equal amplitude traveling in opposite directions and coinciding for an instant at time t = 0. The arrows pointing to the right and to the left indicate the directions of travel of the waves. Their resultant displacement is shown by the dashed curve. In (b) both waves traveling in opposite directions have advanced one eighth of a period and their resultant is shown reduced in amplitude only. In (c), one eighth of a period later, the resultant is instantaneously zero. Each succeeding graph represents the interference picture one eighth of a period later. In (d) and (e) the resultant has the same amplitudes as in (b) and (a), respectively, but reversed 180°. The resultant amplitude in (f) again decreases, reaching a zero value in (g), and then increases again through (h) to a maximum value in the opposite direction in (i). Notice that the resultant wave crosses the horizontal axis always at the same points, which have zero displacement; these are the nodes indicated by N at the bottom of the figure. Midway between any two adjacent nodes the particles of the medium execute simple harmonic motions up and down with the same frequency as the two interfering waves and with a maximum amplitude as shown. These points of maximum amplitude are the antinodes or loops and their positions are indicated by L at the bottom of the figure. We note that the maximum amplitude-changes in adjacent loops are in opposite directions. We can now sum up our observations that the standing wave pattern on a cord stretched between two rigid supports is characterized by nodes at both ends, nodes every half-wavelength from each end, and loops midway between the nodes. It is of special significance that the nodes are spaced half a wavelength apart and the distance between a node and an adjacent loop is a quarter wavelength or

$$\text{Distance between successive nodes} = \lambda/2 \qquad (4.2)$$

$$\text{Distance between an N and adjacent L} = \lambda/4 \qquad (4.3)$$

The relationships in equations 4.2 and 4.3 are quite general for stationary wave patterns. Since the number of nodes on a stretched cord exhibiting standing waves is one more than the number of half-waves between the supports the length of the stretched cord is $(n-1) \lambda/2$ where n is the number of nodes. We shall be applying the above physical facts to the analysis of vibrating strings and vibrating air columns in the next chapter.

Problems

4. 1.   Two sound waves having the same frequency are traveling in phase through the same medium at the same time. One of the waves has an amplitude of 6 units and the other an amplitude of 4 units. On a sheet of graph paper make a plot of the two waves and, using the principle of superposition, obtain a plot of the resultant wave. What is the amplitude of the resultant wave?

4. 2.   When two tuning forks are sounded together 3 beats per second are heard. If one of the tuning forks has a frequency of 237 cycles/sec, what can you say about the frequency of the other fork?

4. 3.   If your answer in Problem 4.2 is not unique describe how you would go about to determine the true frequency of the other fork.

4. 4.   When the sound from a sonorous source is emitted with one standard source of 440 cycles/sec, 6 beats per second are heard and when sounded with another standard source of 450 cycles/sec, 4 beats per second are heard. What is the frequency of the sonorous source?

4. 5.   One end of a cord under tension is fastened to a rigid support
         and the other end is vibrated transversely or perpendicular to
         the cord length at a frequency of 300 cycles/sec. A stationary
         wave pattern is produced with nodes spaced 0.75 meter apart.
         Find the speed of propagation of the transverse wave.
4. 6.   Explain why the waves coming from the two prongs of a
         tuning fork interfere destructively at four positions when the
         fork is rotated as described in Section 4.1.

CHAPTER 5

Vibrating Systems

Before proceeding with descriptions of vibrating strings, vibrating air columns, and other vibrating elements and systems it will be useful to discuss the very important phenomenon of resonance that enters into the description of vibrating systems. We shall therefore devote the next section to a consideration of resonance before proceeding with the subject of this chapter.

5.1. Resonance

We have all, at one time or another, experienced or observed the phenomenon of resonance. Reflect for a moment on the mechanics of the oscillation of a person sitting on the seat of a swing. How do we accomplish increasing the height of the swing i.e. increase the amplitude of vibration? Each time that the person approaches you and is on the verge of swinging forward you gently push the person forward and continue to repeat this pattern. In other words, whenever properly tuned impulses are imparted to a system that is capable of oscillating or vibrating, and the frequency is equal to or is in tune with a characteristic frequency of the system, the amplitude of oscillation or vibration of the system increases. This phenomenon is called *resonance*. Every body or system has one or more characteristic frequencies of oscillation or vibration and when there are present external impulses that match one of these frequencies resonance occurs. The external impulses may either be equal to or be an integral submultiple of this natural frequency of oscillation or vibration of the system. If the frequency of the external impulses is somewhat lower or higher than that of the system, the amplitude of vibration built up in the system is less than when they are equal.

To exhibit the phenomenon of resonance experimentally we can apply a variable excitation frequency to a vibrating system, and observe the amplitude of vibration as the variable frequency sweeps through the natural frequency of oscillation or vibration of the system. This can be done acoustically, mechanically, electrically, etc. By plotting the amplitude of vibration against the driving or exciting frequency f one obtains the well known resonance curves shown in Figure 5.1. The frequency f_r represents the natural or resonant frequency where the amplitude of vibration is maximum. When the exciting frequency is sufficiently removed from the resonant frequency, either above or below f_r, the amplitude of vibration is small. As f approaches f_r the amplitude rises as shown and when it is equal to f_r the maximum response occurs. The curve is assymetric and its height and its sharpness depend upon the amount of damping present due to friction or other form of energy dissipation. When the damping is high the resonance is broad while when the damping is low the resonance is sharp. Theoretically, if there were no damping, the resonant peak would be infinite in height. Some systems, like the simple pendulum whose bob oscillates at the end of a cord, have only one natural frequency which for a pendulum depends solely on its length, and hence only one resonance curve. Other systems like the vibrating string and the vibrating air column have innumerable different natural frequencies and corresponding resonance curves or resonances.

Resonance phenomena are quite common. We have discussed the swing or pendulum system above. Tuning a radio involves adjusting an oscillating electric circuit until its natural frequency matches the frequency of the incoming radio signal; then resonance obtains and the electrical circuit response is a maximum. Soldiers in marching across a bridge purposely break step to prevent properly timed impulses from causing violent vibrations of the bridge and thus avoid a resonant catastrophe. A striking example of mechanical resonance occurred in 1950 when a suspension bridge at Puget Sound in Washington was set into resonant vibration by gale winds resulting in such large amplitudes as to collapse the bridge. We shall later in this chapter see that if the stem of a tuning fork is mounted on a wooden box, closed at one end and open at the other, and of the proper length so as to act as a resonator, the column of air in the box will oscillate sympathetically with the fork, resulting in the emission of a much louder sound from this mass of oscillating air than from the fork itself. Also, when two such forks of the same frequency are mounted on

separate resonator boxes, and one of the forks is made
quickly stopped by touching it, the second fork will pick
and vibrate by resonance or by what is also called
vibration.

In general the sound-producing effectiveness of all musical
instruments depends upon the principle of resonance. We shall now
see how important a role the phenomenon of resonance plays in the
vibrating string, the vibrating air column, and other vibrating
systems.

5.2. The Vibrating String.

In Section 4.3 we summarized that when a string under tension is
actuated to vibrate (bowed, struck, or plucked) a transverse disturbance
travels to both ends where reflections take place, thus forming
stationary waves with displacement nodes at both ends, where the
string is attached to rigid supports, and every half wavelength
therefrom. Figure 5.2 shows three of these nodes N which have
originated from each end. In general, for an arbitrary wavelength, the
system of nodes and loops proceeding from the right will not coincide
in position with those proceeding from the left. However, if the length
of the string is adjusted so that the standing waves formed by reflection
from the right end reinforce those formed from the left end, resonance
occurs and a standing wave pattern along the entire length of the
string will obtain when the two systems of nodes and loops coincide.
This condition is fulfilled in two ways: either the wavelength is such
that an integral number of half-wavelengths fits into a fixed length of
string or the string length is such as to accommodate an integral
number of fixed half-wavelengths. For a vibrating string of a given
length these conditions are automatically fulfilled, and sustained
stationary waves can occur in any one of a series of modes. The first
three members of the different modes of vibration are shown in Figure
5.3. In (a) is shown the standing wave pattern for the simplest mode of
vibration with a displacement node at each end of the string and a loop
midway between them. The string vibrates in one segment, oscillating
up and down and forming the half wavelength pattern. This mode is
called the *fundamental* or *first harmonic*. Here twice the length of the
string, 2l, is equal to the wavelength and the frequency of vibration is
$f_1 = v/\lambda_1 = v/2l$ where v is the speed of the transverse wave and is
fixed for a given string under a given tension. In (b) we have the next
simplest mode which contains three nodes and vibrates in two

segments with adjacent segments oscillating 180° out of phase, that is their motions are at all times opposite in directions with the right-hand loop down while the left-hand loop is up and vice versa. This mode is known as the *second harmonic*, sometimes also called the *first overtone*, and its frequency is $f_2 = v/1 = 2(v/2l)$. In (c) we have another node added, the string vibrates in three segments with adjacent segments 180° out of phase. This mode is known as the *third harmonic* or *second overtone* and its frequency is $f_3 = 3(v/2l)$. The next mode would be the fourth harmonic with four half wavelengths falling between the ends and a frequency of $f_4 = 4(v/2l)$. Each succeeding harmonic corresponds to a standing wave pattern containing an additional node. It is therefore seen that both the even and the odd harmonics are possible in a vibrating string with the frequency of the nth harmonic being given by

$$f_n = \frac{nv}{2l} = nf_1 \quad n = 1, 2, 3,....$$ (5.1)

showing that each harmonic has a frequency that is an integral multiple of the fundamental frequency. Now it proves to be true, and we shall not pursue the derivation, that the speed of the tranverse wave along a stretched string under a tension F and of mass per unit length m_u is given by

$$v = \sqrt{F/m_u}$$ (5.2)

so that by substitution into Equation 5.1 we have

$$f_n = \frac{n}{2l} \sqrt{F/m_u}$$ (5.3)

A string therefore has multitudinous natural frequencies given by Equation 5.3 and when one end of the string is driven by a mechanical vibrator whose frequency coincides with one of the natural frequencies f_n, resonance occurs and we can readily observe large-amplitude standing waves of wavelength v/f_n. This condition of resonance can be achieved by tuning the driving frequency to coincide with f_n, or by varying the length, tension, or mass of the string to adjust the natural frequency of the string f_n to coincide with the driving frequency. Violinists, violists, cellists, etc., tune their instruments by adjusting the tension F of the strings which have a fixed l and m_u. In addition, the string may be shortened and its fundamental frequency increased by pressing the finger on the string against the fingerboard. The different strings on an instrument such as the violin have different values of m_u, those of smaller linear density having a higher fundamental frequency in accordance with Equation 5.3.

As we stated earlier, a medium can carry many waves simultaneously so that a string is capable of vibrating in a number of modes simultaneously, the waves of different wavelengths, as they pass through one another, giving rise to a complex resultant wave shape in the string. Precisely which harmonics are present, and what their relative intensities are, depends upon the kind of excitation employed and upon the point at which the string is excited. The particular harmonics evoked and the distribution of their relative intensities are different from a string that is bowed, plucked or struck. Also when bowing across a string, the speed of the movement of the bow, its point of contact and its pressure have a pronounced effect on the harmonics produced and on their relative intensities. Ordinarily when a string is struck, as on a piano, or when bowed, as on a violin, it is done about one-eighth the effective length of the string from one end in order to minimize or eliminate the seventh and the ninth harmonics which are quite dissonant. To see how the point of excitation on a string has this kind of effect we call attention to the fact that at the point where the string is excited there must be a loop and a node cannot exist, and when a bow is moved across a string it drags the string for a short distance until the restoring force causes the string to slip back past its equilibrium position and the cycle is repeated as the string is bowed in one direction. Returning to Figure 5.3 and considering the patterns of some higher harmonics, we can see that if the string is pulled up at its center point and released the even harmonics will be absent, because they require the presence of a displacement node at the midpoint of the string, and the amplitudes of the odd harmonics will decrease with an increasing number of harmonics. Again, if the string is excited at a point one-fourth of its length from one end, there will be present the second harmonic with a large amplitude, the third harmonic with a very small amplitude; and the fourth harmonic, as well as all integral numbers of the fourth harmonic, will be absent.

It is important to repeat and emphasize that the coexistence of harmonic waves on a string with frequencies that are integral multiples of a fundamental frequency is a significant illustration of the principle of superposition. We have learned that the several coincident disturbances or vibrations give rise at any point to a resultant effect which is obtained graphically by plotting the displacements of the component vibrations with their correct phases as a function of the time, and adding the ordinates for each value of the time. The Fourier theorem, referred to earlier in Section 2.4, enables us to effect the reverse process of analyzing or decomposing a complex wave or vibration into component simple harmonic vibrations. This theorem states that any periodic wave may be represented as the sum of a number of simple harmonic waves with frequencies that are integral multiples of one fundamental frequency. For some complex waves a small number of simple harmonic waves is sufficient to represent the complex wave, for other such waves a large number of simple harmonic components, and even an infinite number, is required for an exact representation.

5.3. The Vibrating Air Column.

An air column will transmit longitudinal waves just as a stretched string will transmit transverse waves. When air or other gas is confined within a rigid pipe or tube of finite length, a sound disturbance produced at one end is propagated along this column of gas as a longitudinal wave. It then reflects from the ends of the tube in a fashion similar to the reflections of transverse waves in a stretched string. These waves, traveling in opposite directions through the gaseous medium, give rise to standing waves. Such vibrating sound-emitting gaseous columns, with the attendant resonant properties, may be of the open type, in which both ends of tube are open, or of the closed type in which one end is open and the other end is closed. Organ pipes and wind instruments provide examples of vibrating air columns that, in like fashion to the string, exhibit a series of harmonic modes of vibration.

To develop these modes of vibration we must first look into the character of reflection that a sound wave experiences when its condensations and rarefactions reflect from the open and closed ends of a pipe and so determine whether a displacement node or displacement loop is formed at these ends. When a longitudinal wave reflects inside a tube at the closed end the condensations and rarefactions, traveling in the acoustically less dense gaseous medium, reflect at the boundary of an acoustically more dense medium that forms the closed end. At the boundary of the more dense kmedium there is a 180° change of phase upon reflection, a condensation, or region of high pressure, returning as a condensation and a rarefaction, or region of low pressure, returning as a rarefaction. This can be better understood if we depict the lower dense medium by a string of elastically connected small spheres and the more dense medium by a string of elastically connected larger more massive spheres. When a compressional pulse is directed from the less dense to the more dense medium by effecting on one of the small balls an impact in the direction toward the more dense medium, the compressional pulse is transmitted from one small ball to the next through the elastic coupling and gets to be reflected as a compressional pulse when the last small ball strikes the first of the larger balls, which have the greater inertia, in the more dense medium. Hence, a condensation returns as a condensation upon reflection at the more dense medium. Likewise, a similar argument leads to the result that a rarefaction, created by pulling on one of the small balls, returns as a rarefaction upon reflection at the more dense medium. Now, remembering that the particles of the medium in a condensation are moving in the same direction of wave travel, the displacement of a particle at the boundary undergoes a 180° change of phase upon reflection and there is therefore found a displacement node at the boundary of the more dense medium.

When a longitudinal wave reflects inside a tube at the open end the condensations and rarefactions reflect with no change in phase consistent with their traveling from an acoustically more dense medium (confined gas) to an acoustically less dense medium (free gas). By going through the exercise of directing a compressional pulse from the more dense to the less dense medium, using the above ball analogy, it should be clear that a condensation returns as a rarefaction and a rarefaction returns as a condensation upon reflection at the boundary of the less dense medium. Again, remembering that the particles in a rarefaction are moving in a direction opposite to that of wave travel, there is always formed a displacement loop or antinode at the open end of the tube. The reader may wish to think of the situation in this way: when a condensation reaches the open end it expands from a confined region in the tube into an open region outside of the tube and a rarefaction reflects back into the pipe; or at an open end of a pipe containing air as the vibrating medium the pressure is the surrounding air pressure so the amplitude of the pressure vibration must be theoretically zero and there would be a pressure node or a displacement loop at the open end (see Section 3.1 and Problem 3.4).

We are now in a position to discuss the modes of vibration of open and closed air columns shown in Figure 5.4. First consider the open air column of length l_o in (a). The simplest mode of vibration is the fundamental or first harmonic with a displacement loop L at both ends and a displacement node N in the middle. Here the tube length l_o equals half a wavelength and the frequency is $f_1 = v/2l_o$ where v is the speed of sound in the tube. The next member of the series is the second harmonic with an additional loop added at the center. Here the length of the tube is equal to one wavelength and the frequency is $f_2 = v/l_o = 2(v/2l_o)$ or twice the fundamental frequency. Proceeding to the third harmonic, for which there is added another displacement loop, we have for the frequency $f_3 = 3(v/2l_o)$ or three times the fundamental. Succeeding modes of vibration have frequencies that are even and odd integral multiples of the fundamental, as in the vibrating string, both the even and the odd harmonics being permissible.

In Figure 5.4b are shown the first three members of the closed type tube of length l_c. In its simplest mode, the fundamental or first harmonic, there is a displacement node N at the closed end and a displacement loop L at the open end. Here the length of the tube is equal to a quarter wavelength and the frequency is $f_1 = v/4l_c$. The second member of this series contains an additional displacement node and an additional displacement loop L. For this mode the length of the tube is three-quarter wavelength of $l_c = 3\lambda/4$ and the frequency is three times that of the fundamental or $f_3 = 3(v/4l_c)$. Therefore this member is the third harmonic or first overtone. The third member of the series contains three displacement nodes, $l_c = 5\lambda/4$ and $f_5 = 5(v/4l_c)$. This is called the fifth harmonic or second overtone. Succeeding modes of vibration have frequencies that are odd integral multiples of the fundamental frequency so that only the odd harmonics are possible, the even harmonics being physically inadmissible.

In Figure 5.4 the standing wave patterns for each mode have been drawn in and the integral subscript on the symbol f conveniently equals the number of the harmonic. If we compare the frequencies for the open and closed pipes, it is clear that odd harmonic frequencies of an open pipe of a given length can be obtained with a closed pipe of half the length. The closed-pipe series of harmonics is termed a quarter-wave system while those of the open pipe, as well as of a vibrating string, correspond to a half-wave system. As in the case of a string, when a column of air vibrates the fundamental and the harmonics are emitted at the same time.

In our discussion we assumed that the displacement loop L occurs exactly at the open end of a pipe. This is not strictly true because the air is moving in and out of the open end and there is an expansion of the disturbance into the region beyond the open end. This makes the tube appear longer than it actually is and the loop L appears a short distance beyond the open end by an amount called the end correction. For a cylindrical pipe whose radius r is small in comparison with the wavelength, the end-correction distance that must be added to the open end to locate the center of the displacement loop is approximately 0.6r. If the cylindrical pipe is open at both ends, twice the end correction must be added to obtain the effective length of the pipe.

5.4. Other Vibrating Systems: Rods, Plates, Membranes, etc.

A metal rod may be set into longitudinal vibration by clamping it at some point and stroking it lengthwise with a rosined cloth. The mode of vibration that is produced depends upon the location of the clamp where a displacement node is physically made to exist. Figure 5.5 shows some of the modes of longitudinal vibrations of a rod clamped at different positions C. In (a) the rod is clamped at one end and its fundamental mode of vibration has a displacement node at the clamped end and a loop at the free end. Here the length l of the rod is one-quarter wavelength so that $\lambda = 4l$ and the frequency is $v/4l$. In (b) is shown the next mode of vibration with an additional node at one-third the length of the rod from its free end so that $l = 3\lambda/4$ and the frequency is $3(v/4l)$ which is the third harmonic or first overtone. In (c) the rod accommodates an additional node as shown, $l = 5\lambda/4$, the frequency is $5(v/4l)$ and we have the fifth harmonic or second overtone. Thus the rod clamped at one end produces a series of standing wave patterns and corresponding vibrational modes that resemble those of the closed pipe with the existence of only the odd harmonics. In (d) there is shown the standing-wave pattern for the fundamental mode of the longitudinal vibration of a rod clamped at its midpoint where a node exists with loops at the two free ends. Here l = $\lambda/2$ and the frequency is $v/2l$. In (e) and (f) are the next two modes of vibration of a rod clamped at its center with frequencies $3(v/2l)$ and $5(v/2l)$; note that here the succeeding modes are obtained by adding nodes in pairs symmetricaly positioned on opposite sides of C. In (g) the rod is clamped at a point one-fourth its length from one end, or from both ends as in (h). Here the standing-wave pattern corresponds to a frequency v/l or $2(v/2l)$ which is the second-harmonic of the fundamental when the rod is clamped at its midpoint.

Not all oscillating systems that possess more than one mode of vibration will have higher mode frequencies that are integral multiples of the fundamental frequency. For example *inharmonic* overtones are evoked by distorted air columns, by resonant tubes containing holes drilled into its side and by the transverse vibrations of metal or wood bars or rods. Rods and bars may be made to vibrate transversély if they are clamped and struck sharply and for these the frequencies are inharmonic; i.e. their overtones are not integral multiples of the fundamental frequency. Figure 5.6a depicts the modes of transverse vibrations of a bar clamped at one end and free at the other end. Here the first, second, and third overtones have frequencies that are respectively 6.27, 17.55 and 34.40 times that of the fundamental. Musical instruments that have as their vibrating element a bar or reed clamped at one end are the harmonica, the clarinet, the reed organ, the oboe, the bassoon and the saxophone.

In Figure 5.6b are shown the transverse vibrational modes of a bar of metal or wood supported at two locations. Here the inharmonic frequencies of the first, second, and third overtones are respectively 2.76, 5.40 and 8.93 times the frequency of the fundamental. For the fundamental the nodes are distant 0.2241 from the free ends. Musical instruments that employ such a free bar are the marimba, the xylophone, the glockenspiel, and the chimes or tubular bells.

The tuning fork may be considered as a transversely vibrating bar supported at two locations as in the fundamental of Figure 5.6b but bent to form a U and having a short stem extending below the curved base. The two nodal points are thus brought closer together and are located on either side of the stem as indicated in Figure 4.2. If we concentrate on the transverse vibrations of one of the prongs we may consider the vibration as that of a bar fixed at one end and free at the other end as inFigure 5.6a. When one of the prongs is struck, a transverse vibration is set up, the prongs vibrating toward or away from each other. After striking a tuning fork, any overtones that are present are transient and quickly dissipate leaving a tone that persists at a single frequency. Tuning forks are therefore often conveniently employed as frequency standards. The output of a tuning fork may be increased by mounting the base of the fork upon a wooden resonant box containing an air column whose resonant frequency is tuned to the frequency of the fork. In this way the vibration at the base of the fork is transmitted to the box whose vibrating walls act as effective couplers to set the air column into vibration for a much greater sound intensity. It should at all times be kept in mind that a mechanically vibrating body, like the rod, bar or reed, whether transverse or longitudinal, always produces a compressional or longitudinal wave in the air or other medium.

Our discussion thus far has been confined to one-dimensional oscillators. When a thin metal plate is clamped at its center and excited by striking it or bowing it on its edge, it will vibrate in a complicated two dimensional fashion. Likewise, when a plate or membrane is supported at its outer edge and struck, the plate or membrane will exhibit a series of vibrating segments, analogous to the cases of vibrating strings, rods and bars. In these two-dimensional vibrators there are present two-dimensional waves that reflect from the boundaries resulting in standing wave patterns that are characterized by nodal lines rather than points. These nodal-line patterns may be demonstrated by supporting a plate or membrane horizontally, sprinkling fine sand on the surface, and causing it to vibrate by bowing on its edge. The sand particles readily move away from the regions of large displacement and accumulate on the nodal lines where there is least motion. A great variety of vibrational modes is possible depending on the manner in which the two-dimensional vibrator is supported and on the means of excitation. For example, for a circular membrane or drumhead clamped around its circular edge, the circular edge is a nodal line or locus and the fundamental vibrational mode coressponds to the surface of the membrane oscillating as a whole with the center of the drumhead having the maximum displacement in a direction perpendicular to its plane. The overtones have nodal lines that are concentric circles, diameters or combinations of these and with frequencies that are inharmonic. The first overtone is characterized by a diametral nodal line in addition to the circular nodal line around the edge. This forms two semicircular sections that oscillate with opposite displacements above and below the equilibrium plane of the drumhead with a frequency 1.593 times that of the fundamental. The second overtone has two diametral nodal lines at right angles, thus dividing the surface into four equal sector segments. The adjacent segments have opposite displacements and the frequency is 2.136 times that of the fundamental.

We can now summarize our discussion of vibrating systems by stating that all elastic bodies or systems vibrate with characteristic frequencies that correspond to given boundary conditions, such as the end structures of open or closed tubes, or the manner in which rods or bars are clamped, or the circular boundary of a stretched membrane. In some systems the frequencies are related harmonically, while in other systems the frequencies are inharmonic and are therefore not integral multiples of the fundamental frequency. However, in all cases there are standing waves characterized by nodal points in one dimensional vibrations, nodal lines in two-dimensional vibrators, and, we trust the student can by extrapolation agree, nodal surfaces in three-dimensional vibrators.

Problems

5. 1.   Describe the difference between sharp resonance and broad resonance and give examples of each type.

5. 2.   Describe the application of resonance in the piano and the violin.

5. 3.   Discuss how the principle of resonance applies in the emission of sound from an oscillating air column.

5. 4.   A cord is 1.05 meters long and has a mass of 0.350 gram. If the string is under a tension of 24.5 newtons, with what speed will a transverse wave travel along the string?

5. 5.   A string 93.0 cm in length and of total mass 0.465 gm is under tension. One end of the string is fixed and the other end, attached to a vibrator, vibrates at 120 Hz transverse to the length of the string. What must be the tension if the standing-wave pattern contains four nodes?

5. 6.   A string under tension is vibrating in two segments. If only the tension is quadrupled, into how many segments will the string vibrate? Answer the same question if only the tension is decreased to one-fourth its original value.

5. 7.   One prong of a tuning fork, vibrating at 100 Hz, is attached to one end of a cord under tension so as to vibrate the cord transversely. The cord has a mass per unit length of 3.60×10^{-5} kg/m, is under a tension of 25.0 newtons and is vibrating in its fundamental mode. Find the length of the cord.

5. 8.   Do Problem 5.7 under the condition that the prong is attached to the cord so that its line of vibration is along the length of the string. (Hint: In this case the frequency of vibration of the string is just half the frequency of the fork. Explain why.)

5. 9.   Design an experiment and indicate what data you would observe to prove the dependence of the frequency f on the tension F and on the mass per unit length m_u as given by Equation 5.3.

5.10.   Referring to Figure 5.3 consider pulling the string up at a point one-third the length of the string and then releasing it.

What harmonics would be present and which would be absent?

5.11. Using the model of more massive and less massive balls to represent the elastically connected particles respectively of a more and a less acoustically dense medium, show by diagrams and explanations that when a series of condensations and rarefactions are reflected from an acoustically more dense medium a displacement node is formed at the reflecting surface. Also show that if the reflection takes place at the boundary of an acoustically less dense medium an antinode or loop is formed at the reflecting surface.

5.12 A vertical glass tube open at the top and fitted with a constrictive valve at the bottom is filled with water. When a tuning fork vibrating at 512 Hz is held above the tube and the water is permitted to discharge slowly through the valve, resonances are detected. The first resonance in the air column occurs when the water level is 15.7 cm from the top, and the next resonance occurs 48.1 cm from the top. Find (a) the speed of sound in the air column and (b) the end correction.

5.13. The third harmonic of an open pipe has the same frequency as the fifth harmonic of a closed pipe whose length is 75.0 cm. What is the length of the open pipe? Sketch the standing wave patterns for each pipe. The vibrating medium in both pipes is air at the same temperature.

5.14. The apparatus shown in Figure 5.7 is used to measure the speed of sound in a gas or in a metal rod by the well-known Kundt method. A metal rod of length l is clamped at its midpoint as shown, with one of its free ends attached to a stopper S. The rod is free to vibrate longitudinally in a glass cylindrical tube G. At the other end of the resonant tube is an adjustable plunger P. When longitudinal vibrations are set up in the rod by stroking it with a rosined cloth, the air column is adjusted in length by P until resonance occurs. The standing-wave nodes and loops are detected by lycopodium powder or cork dust which has previously been spread throughout the interior of the tube. At resonance the powder becomes visibly agitated and its particles line up in detectable displacement node and loop patterns. Given that the rod is 100 cm long and the observed average distance d between displacement loops in the dust pattern is 6.80 cm, find (a) the speed of sound in the rod and (b) the frequency of vibration of the rod. The gas in the tube is air at 20.0° C. (Hint: the frequency of vibration in the rod and in the air is the same; the speed of sound in the air at 0° C is 332 m/sec.)

5.15. A tuning fork with a frequency of 440 Hz is mounted on a wooden box to increase the loudness. This resonant box is closed at one end and open at the other end. If the end correction is 0.8 cm, how long should the box be so that the resonance is most effective? Take the speed of sound in air as 340 m/sec.

5.16. Two closed organ pipes are vibrating in air in their fundamental modes. They produce 30.0 beats in 15.0 sec. Neglecting end corrections, find the length of the longer pipe if that of the shorter is 75.0 cm. Take 332 m/sec as the speed of sound in air.

5.17. How do the longitudinal modes of vibration of a rod clamped at one end differ from those of the same rod clamped in the middle?

5.18. Besides stroking a rod with rosined cloth how else could the rod be set in longitudinal vibration?

CHAPTER 6

Subjective and Objective Characteristics of Sound

The human receptor of sound is the ear which is actually a comparatively complicated, but a beautifully and efficiently constructed instrument whose mechanical and physiological action make the ear perceive musical sounds that differ from one another in three ways which are designated as the subjective characteristics known as *pitch, loudness* and *quality*. In this chapter we consider and discuss these in turn but we shall first address ourselves briefly to a description and action of the human ear.

6.1. Structure and Action of the Ear

Instead of referring our discussion to an accurate reproductive diagram of the ear, which may be found in other numerous scientific and anatomical publications, it is more instructive for our purposes to employ the schematic diagram shown in Figure 6.1 which is not drawn to scale. The ear is divided into three sections which are named the *outer ear, the middle ear* and the *inner ear*. In the outer ear there is the *pinna* or *trumpet* which serves to collect and funnel the sound into the *auditory canal* or *meatus* which is about 2.7 cm in length and about 0.7 cm in diameter, with one end open and the other end closed by the *eardrum* or *tympanic membrane*.

The meatus is essentially a closed organ pipe that sets up stationary sound waves of wavelength about 10.8 cm. At the normal temperature of the body, 37° C, the speed of the sound waves in the meatus is 355 m/sec, giving for the fundamental resonant frequency 355/0.108 or about 3287 Hz. We have here neglected the conical shape of the meatus in this computation. With some amplification provided by the meatus the ear exhibits the greatest sensitivity to frequencies between about 2000 Hz and 5000 Hz.

On the other side of the eardrum is the region of the middle ear which contains a series of three small boney ossicles called the *hammer*, the *anvil* and the *stirrup* which are connected but not rigidly. The ossicles serve, by mechanical lever action, to transfer the vibrational energy of the eardrum to the inner ear, the hammer being attached at one end to the eardrum and the stirrup being attached to the membrane called the oval window at the entrance to the inner ear. The lever system effects an increase in force which, acting on the small area of the oval window relative to that on the eardrum, results in a pressure on the oval window about fifteen times that on the eardrum. The middle ear region also connects to the back of the throat by the *Eustacian tube* which permits the equalization of the changes in ambient air pressure thus preventing excessive distortions of the eardrum. Normally the Eustachian tube seals the middle ear cavity and it may be opened by swallowing or yawning. This permits air to enter or leave the region and equalizes the pressure on both sides of the eardrum. The middle ear in its action is non-linear; consequently vibratory displacements are not transmitted in a proportional manner. This results in harmonic distortion so that if the input to the ear drum contains a single frequency f_1, then there is a doubling, $2f_1$, tripling $3f_1$, etc, in the output in addition to f_1; or if there are two frequencies f_1 and f_2 in the input, the output additionally contains difference and sum frequencies such as $f_1 + f_2$, $f_1 - f_2$, $2f_1 - 2f_2$, $2f_1 + 2f_2$, etc.

The inner ear consists in part of a cavity called the *cochlea* which is a tube with a slight taper and coiled like a snail shell. It has an average diameter of about 1.5 mm, makes about 2¾ turns, is divided down its length by a soft partition called the *basilar membrane*, and filled with a *perilymph* fluid. In Figure 6.1 the spiral cochlea is shown uncoiled to facilitate our discussion. The basilar membrane divides the cochlea into two chambers; one chamber, the *upper gallery*, is separated from the middle ear by the *oval window* which is attached to the stirrup; the other chamber, the *lower gallery*, is separated from the middle ear by the *round window*. Both chambers communicate by a small opening, of area about $0.3 mm^2$, called the *helicotrema*. Embedded in the basilar membrane are thousands of nerve endings which sense the sound disturbances and transmit them to the brain. The cochlea is the portion of the inner ear that possesses the hearing function. The other portion of the inner ear contains the *vestibular process* which is anatomically adjacent but physiologically independent of the cochlea. The vestibular process consists of three semi-circular canals which are perpendicular to one another and serve to orient a person as to up and down and maintain directional equilibrium. The semicircular canals are not involved in the hearing process.

When sound disturbances pass down the auditory canal to the eardrum they cause the eardrum to vibrate. This vibration is transmitted through the ossicles to the oval window of the cochlea and the accompanying pressure is transmitted to the cochlear fluid. To permit the fluid to move, since fluids are relatively incompressible, the round window is constructed to move outward as the oval window moves inward. This fluid motion is communicated to the basilar membrane in whose surface there terminates the thousands of auditory nerve endings which act as sensors thus resulting in stimuli that are transmitted to the brain. The basilar membrane acts as a kind of wide band mechanical filter that separates a sound containing many frequencies into its components with the result that certain groups of nerves are excited much more by one frequency than by any other frequencies, resulting in a given sense of pitch which is related to vibrational frequency as explained in the next section.

6.2. Pitch and Frequency

When the frequencies of sound-emitting vibrating elements are measured it becomes readily apparent that the subjective characteristic of *pitch* is closely related to the physical quantity frequency. Low pitches correspond to low frequencies and high pitches to high frequencies but pitch and frequency are not proportional to one another. Whereas frequency is a definite quantity measurable by an instrument in cycles per second or hertz, pitch is not only a function of the frequency, but is also dependent upon the sound pressure level or loudness and on the complexity of the sound. Figure 6.2 shows the non-linear relationship between the pitch, as judged by an average observer and designated in a unit called the *mel*, as a function of the frequency. Such a curve is obtained by experimenting with observers who are asked to adjust the frequency of an audio oscillator connected to a loudspeaker until it is judged to have twice the pitch of a second reference oscillator that is alternately connected to the loudspeaker. Such a procedure is repeated for a range of frequency settings of the reference oscillator. The curve has very nearly the same shape as the curve that relates the positions of the auditory nerve groups, along the basilar membrane, that are stimulated as a function of the frequency. Thus the judgment of pitch appears to be associated with the location of points of excitation of a given frequency along the basilar membrane. In constructing such a curve the average of the judgments of many observers is used and a reference pitch of 1000 mels is chosen as the pitch of a 1000 Hz tone at a given sound pressure level.

It will become evident (see Figure 6.4) that as we go to lower frequencies the ear becomes less sensitive and, as shown in Figure 6.2, below some critical frequency f_c the ear does not respond to the sound vibrations. This critical frequency for audible sounds varies greatly from person to person but a rough value for the lower frequency limit is about 16 cycles per second. At the other end of the frequency scale there is an upper limit to the highest frequency that the ear can hear and this limiting frequency also varies from one person to another and changes with age. For example, a young person can hear sounds of frequencies as high as 18,000 cycles per second and even somewhat higher; in an older person this upper limit may drop to 12,000 or 10,000 cycles per second. As a rough value for the upper frequency limit we may take about 16000 cycles per second. Thus, the ear is capable of hearing a frequency range of about 1000 to 1. By contrast, the frequency range of the human eye for visible light is only about 2 to 1.

The pitch of a sound of a given frequency. depends on several factors which may be responsible for the fact that a specific frequency will not always be perceived as the same pitch. For one thing, pitch depends somewhat on the loudness or intensity of a pure tone, and what is subjectively perceived may be an increase or a decrease as the loudness is increased, depending upon the frequency of the tone. For example, for pure tone frequencies between about 200 Hz and 1000 Hz the pitch decreases as the loudness increases and for pure tone frequencies above about 4000 Hz the perceived pitch increases with increase in loudness. For pure tone frequencies between about 1000 Hz and 4000 Hz changes in loudness appear to have very little effect on pitch. Such subjective judgments of pitch are not experienced in the same way by different persons and the effects are for pure tones and not for complex tones. The tones emitted by musical insktruments are in the main complex in structure and for these there appears to be no change in pitch with loudness.

Another factor that affects the perception of pitch is the length of time that a sound lasts; if the sound lasts for only a small fraction of a second, say 0.01 sec, and the vibrational frequency is 50 Hz, there is only half of a complete oscillation, the sound is perceived as a click and it is doubtful whether there can be assigned a definite pitch. If the frequency of vibration is 1000 Hz then there are ten complete oscillations in 0.01 sec and this is sufficient to produce a sensation of pitch. Therefore, depending on the vibrational frequency, a sound must have a minimum length of time in order for it to have an identifiable pitch. Musically this effect on pitch perception is of no great consequence since musical tones usually last sufficiently long to produce the sensation of pitch.

We have stated in the preceding section that when there are two or more frequencies in the sound the ear introduces the sensation of the sum and difference frequencies. Thus when a complex tone consisting of the frequencies 200, 300, and 400 Hz is emitted most people will hear the same pitch as that produced by the frequency 100 Hz. The ear has the sensation of the pitch corresponding to the fundamental frequency even though there is no physical vibration at the fundamental frequency. The pitches that the ear perceives at multiples, differences and sums of frequencies that are not present in the complex tone are called *subjective tones* or *Tartini sounds*, so named after the great Italian violinist Giuseppi Tartini (1692-1770) who presumably first recognized these different tones. Such subjective tones, generated by the nonlinearity of the ear, are also called *aural harmonics*.

Thus far we have made no mention of how the relative motion of a source of sound and the ear affects the observed pitch. We are all familiar with the observation that the sounding horn on an automobile appears higher in pitch when the automobile is approaching a stationary observer and appears lower in pitch when the automobile is receding from the observer. This same phenomenon occurs when the source is stationary and the observer is moving toward or away from the source. This apparent change in pitch as sensed by the ear, when there is relative motion between source and observer, is known as the *Doppler effect*. We can explain this as follows: When the observer is moving toward a stationary source the ear receives more waves each second than when the observer is at rest, with a consequent apparent increase in observed pitch. If the observer is receding from the stationary source the ear receives fewer waves each second than when the observer is at rest and there is an apparent decrease in the pitch as heard. When the source of sound is moving toward a stationary observer there is an apparent decrease in the wavelength with a consequent apparent increase in the pitch as heard by the observer. If the sound source is receding from the stationary observer there is an apparent lengthening of the wavelength and there is an apparent decrease in the observed pitch. The magnitude of this apparent change in pitch depends upon the relative speed of the source of sound and the observer. The Doppler effect does not have any special musical significance. A small but discernable increase and decrease in pitch vibrato manifests itself when a violinist sways rapidly when playing. Such pitch vibratos are sometimes purposely introduced mechanically by having rotating sound sources such as speakers, vanes, etc.

In discussing the factors that affect pitch we should mention *aural* or *pitch acuity* although this does not really affect the pitch value but rather the degree to which a person can discriminate between pitches that are closer together. Experiment indicates that it is easier to detect differences in pitch for higher frequencies than it is for lower frequencies. With the ear being capable of detecting a change of about three cycles per second in the neighborhood of 1000 cycles per second, the discrimination is around 0.3 percent or about 0.052 of an equally tempered semitone (the equivalence in semitones is covered in Section 7.6). On the other hand, at a lower frequency of say 30 cycles per second a change of three cycles per second amounts to 10 percent which is equivalent to about 1.7 semitones. At frequencies between about 1000 Hz and 5000 Hz the discrimination in pitch remains fairly constant and quite good at somewhat less than 0.3 percent. For higher frequencies the pitch discrimination becomes very poor as with the low frequencies. Above about 12 kilocycles/sec the pitch discrimination is very nearly zero.

The frequency ranges of the piano, various musical instruments, and the voice are shown in Figure 6.3. The piano keyboard frequencies range from 27.5 Hz to 4186 Hz. These and the indicated frequency ranges for the voice and the musical instruments are those of the fundamental frequencies. The overtones that are emitted with the fundamental are greater than the limit of hearing, around 20000 Hz. Not shown on the chart is the range of the organ extending from a low frequency of 16.4 Hz emitted by a 32-foot open pipe, to a high of 8372 Hz, emitted by a ¾-inch open pipe. Although there are always exceptional voices with greater ranges than the average values indicated, the average singing voice has a range of about two octaves. Note also that the highest fundamental frequency in orchestral music is attained by the piccolo and the lowest is reached by the harp with the bass viol attaining the lowest frequency among the common instruments.

Frequencies are measurable to a high degree of precision with modern electronic instruments such as audio oscillators, oscilloscopes, stroboscopic devices, and by comparison with standard frequency signals operated by the U.S. National Bureau of Standards. Descriptions of these instruments and the methods of measurement may be found in other published works. The author has found that assignments for students to write up and report on the function of one or more of these instruments and their use in the measurement of frequency are desirable and very gainful.

Historically the value for the standard musical pitch has varied in different countries from 370 Hz to 567 Hz. However in 1939 the International Pitch Conference held in London adopted 440 Hz as the standard pitch for A above middle C. This is the standard pitch generally employed by orchestras in the United States.

6.3. Loudness and Intensity.

The subjective characteristic of sound known as *loudness* has to do with the magnitude of the auditory sensation. Since loudness is a subjective characteristic it is, like pitch, psychological in character, so that it is difficult to establish an exact quantitative measurement of it. However it is desirable to establish a numerical scale of loudness. The physical quantity that is associated with loudness is intensity which is purely objective depending entirely on the properties of the sound wave and not upon the ear. It is true that the more intense the sound is the louder it is but there is no simple direct proportionality that exists between loudness and intensity. In other words, a given percentage increase or decrease in intensity is not judged as the same percentage increase or decrease in loudness. Furthermore, loudness also depends upon frequency, the sensitivity of the ear being different in the various audiofrequency ranges.

To proceed with the subject of this section we must first discuss the physical quantity intensity. The *intensity* of a sound wave, at any place, is defined as the amount of sound energy that flows, at that place, per unit time through each unit of area that is perpendicular to the direction of wave propagation. This time rate of energy flow per unit area may be expressed in ergs per second per square centimeter or joules per second square meter or watts per square meter, the latter unit showing that we are dealing with acoustic power. It can be shown that the intensity I of a given sound at any point in a medium is given by

$$I = 2\pi^2 \rho v f^2 A^2 \qquad (6.1)$$

where ρ is the density of the medium, v is the speed of sound, f is the frequency and A is the amplitude of the wave disturbance. The significance of Equation 6.1 is that for a given medium the intensity of the sound varies jointly as the square of the frequency and the square of the amplitude. The amplitude itself depends upon the amplitude of vibration of the sound source, the distance of the wave disturbance from the sound source, and the geometric nature of the sound source. For a sound source that acts like a point source the intensity decreases inversely as the square of the distance from the source. In practice the measurement of sound intensity is better accomplished by detecting the pressure variation of the sound wave which is what most microphones measure. In terms of this pressure change the intensity is given by

$$I = p^2/\rho v \qquad (6.2)$$

where p is the effective sound pressure change and ρ and v have the same meaning as in Equation 6.1. If p is in dynes per square centimeter, ρ is in grams per cubic centimeter and v is in centimeters per second, then I is in ergs per sec per square centimeter. We note that the intensity varies as the square of the pressure change.

The human ear is capable of responding to an extremely wide range of intensities. At a frequency of 1000 Hz a sound intensity of 10^{-16} watts/cm^2 is roughly the intensity of the *threshold of audibility*, that is, a sound intensity that can just barely be heard. At the same frequency a sound intensity of 10^{-4} watts/cm^2 is the maximum sound intensity that the ear can tolerate, before physiological pain sets in, and is called the *threshold of feeling*. The ratio is $10^{-4}/10^{-16} = 10^{12}$ which indicates a tremendously wide range of sensitivity. To represent such a range of values it is convenient to use a condensed scale of smaller numbers and the powers of ten suggest that we use a scale of logarithms. Furthermore there is an approximate relationship known as the Weber-Fechner law which states that the response of a human sense organ is proportional to the logarithm of the magnitude of the stimulus. This is the basis of the often stated fact that the ear is a logarithmic device with the loudness of most sounds being approximately proportional to the logarithm of the intensity. Accordingly a decibel (abbreviated dB) scale of sound *intensity levels* has been devised for the comparison of intensities and defined by the expression

$$SIL = 10 \log I/I_o \qquad (6.3)$$

where SIL stands for the sound intensity level in decibels,* I is the intensity of the sound under investigation and I_o is an arbitrary reference intensity usually taken to be on the threshold of hearing at 1000 Hz or 10^{-16} watts/cm^2 with which I is to be compared. It may be recalled that a logarithm of a number is defined as the exponent or power to which a given base (another number) must be raised to obtain the number whose logarithm is being sought. For example, if the base is 10, as in Equation 6.3, then since $10^2 = 100$, the log 100 = 2 or since $10^5 = 100,000$, the log 100,000 = log 10^5 = 5, etc. Returning to Equation 6.3 we may restate it with the value of I_o inserted so that

$$SIL = 10 \log I/10^{-16} = 10 \log (10^{16}I) \qquad (6.4)$$

Now corresponding to the threshold of audibility the SIL = $10 \log 10^{16}$ x 10^{-16} = 10 log 1 = 0 dB and corresponding to the threshold of feeling SIL = $10 \log (10^{16} \times 10^{-4})$ = $10 \log 10^{12}$ = 120 dB and we have a much more convenient and representative scale of sound intensity levels that range from 0 to 120 dB. Table 6.1 lists sound intensity levels of some common sources of sound. A sound intensity level of 1 dB is just barely noticeable and a difference of 1 dB between two sounds is what can on the average just be discerned. By setting SIL = 1dB in Equation 6.3 and

Table 6.1

Sound Intensity Levels of Some Sources of Sound

Sound Source	SIL in dB
Whisper at about 150cm	10
Quiet street, no traffic	30
Average office	45
Ordinary conversation at about 100 cm	70
Noisy factory	85
Heavy traffic	90
Airplane engine	110
Very loud thunder	115
Average pain threshold	120

---

*The unit of intensity level is the *bel*, after Alexander Graham Bell, but this unit is too large and the more usual unit employed is the decibel which is one tenth of the bel.

solving for the ratio of the intensities we obtain I/I_o = 1.26 so that a
1 dB increase in the sound intensity level corresponds to a 26% increase
in sound intensity.

We may use Equation 6.3 to obtain another expression that is useful
in finding the difference between two sound intensity levels. Thus
calling $(SIL)_1$ the sound intensity level corresponding to an intensity I_1
and $(SIL)_2$ that corresponding to an intensity I_2 we have for the
difference

$$(SIL)_1 - (SIL)_2 = 10 \, (\log I_1/I_0 - \log I_2/I_0)$$

Now the difference between two logarithms is the logarithm of their
quotient so that

$$(SIL)_1 - (SIL)_2 = 10 \, \log I_1/I_2 \tag{6.5}$$

As an example of the use of this expression, suppose the intensity of
one sound I_1 is twice the intensity of another sound I_2 then

$$(SIL)_1 - (SIL)_2 = 10 \, \log 2 = 10 \, (0.30) = 3.0 \text{ dB}$$

Therefore, doubling the intensity of sound results in an increase in the
sound intensity level by 3 dB. In the same way if the intensity is
increased by a factor of 10 the SIL is increased by 10 dB and if the
intensity increase is 100 times, the increase in SIL is 20 dB.

From Equation 6.2 we see that we may also specify sound pressures
in terms of decibels. Since the intensity is proportional to the square of
pressure, the *sound pressure level* (SPL) is defined as

$$SPL = 20 \, \log P/P_o \tag{6.6}$$

where P is the sound pressure and P_o is the sound pressure
corresponding to the intensity audibility threshold and has the value 2
x 10^{-4} dyne/cm^2. We may therefore for pure tones obtain the decibel
level of a sound by using either Equation 6.3 or the equivalent
Equation 6.6. In an auditorium, where there are many reflections of
the sound from the walls and other internal structural supports and
overhangs, the SIL and SPL may not be the same. However for our
purposes we need not be concerned about the difference which is
small.

The loudness of a tone is a subjective measurement and depends on
the characteristics of an individual's ears. As many tests have shown,
the loudness level depends on both the sound intensity level, or the
sound pressure level, and on the frequency. By measuring the intensity
or the pressure at which different pure tones of different frequencies
are equally loud it is possible to construct a scale of equal-loudness

contours for the different frequencies of pure tones. Such a series of loudness-level curves, based on the average results of such tests, and adopted by the American Standards Association in 1936, are shown in Figure 6.4. The ordinate is labeled sound intensity level in dB but could just as well have been labeled sound pressure level in dB. The abcissa, on a logarithmic scale, gives the frequencies in Hz. For these curves a pure tone of 1000 Hz with an intensity of 10^{-16} watt/cm^2, or a pressure of 2 x 10^{-4} dyne/cm^2, is taken as the comparison reference value. Now equal loudness judgments occur at different intensity or pressure levels so that a loudness-level unit called the *phon* has been chosen and defined to be numerically the same as the sound intensity level, or sound pressure level, at 1000 Hz. The loudness level of zero phons is a changing threshold of audibility level throughout the entire audible frequency range as indicated by the lowest curve. This curve shows that at the lower frequencies the ear is less sensitive; at 100 Hz the sound intensity level must be nearly 40 dB to be barely heard. Near 1000 Hz the threshold curve drops, indicating an increase in the sensitivity of the ear which is most sensitive in the region between about 3000 Hz and 4000 Hz. We have seen that the auditory canal is essentially a closed pipe with a resonant frequency at about 3287 Hz. As the frequency rises above 4000 Hz the sensitivity of the ear again decreases. The curve directly above the threshold curve is the contour for an equal loudness of 10 phons, which is again set equal to the sound intensity level of 10 dB. From this curve we see that a tone of 100 Hz must have a sound intensity level of about 44 dB to give the same sensation of loudness as a 10 dB tone at 1000 Hz. Examining the loudness level curves at the higher phon values we see that at very high loudness levels the ear is more uniformly sensitive to all frequencies. The uppermost contour, labeled 120 phons, represents the loudness level for the auditory frequency range at which the ear experiences feeling rather than hearing, with the sensation of tickling or even pain. This is the contour of the threshold of feeling.

Now what do the curves in Figure 6.4 really tell us? They simply provide on a relative basis the response or sensitivity of the ear as a function of the frequency for a series of given levels of loudness. They do not give the relationship between the intensity or sound intensity level and the subjective loudness. In other words the curves do not indicate how many phons increase in loudness level result in a sound generally judged to be of twice the subjective loudness. Experimental research* with human subjects who compared sounds and judged

*S.S. Stevens, "Measurement of Loudness," Jour. Acoust. Soc. Am., 27, 815, 1955.

when one sound was twice as loud, three times as loud, half as loud, etc., as another has led to a loudness scale that is representative of the typical listener. The unit of loudness in the scale is the *sone* where one sone is arbitrarily indicated as the subjective loudness of a loudness level of 40 phons at any frequency. Accordingly, two sones are judged twice as loud as one sone, ten sones twice as loud as five sones, etc. A plot of the empirical results yields a linear relationship between the logarithm of the loudness in sones as a function of the loudness level in phons. The results indicate that an increase in loudness level by about 10 phons produces a sound generally judged to be twice the loudness and this relation holds fairly well over the entire range of audible intensities. It must be kept in mind that the phon and sone are subjective measures which are not the same for all listeners and the application to individual cases must be made with caution. It is helpful to reiterate the following: the phon is a subjective measure of a loudness level and is used to describe sounds that are judged as equally loud; the sone is a subjective measure of loudness and is used to relate sounds judged to have different loudnesses; the objective measure of sound is the sound intensity level measured in dB.

As a final topic of discussion in this section it is important to indicate the phenomenon of masking. We have all experienced a sound, which is heard clearly in a quiet room but becomes inaudible in a noisy room, or the experience of being in a noisy environment making it impossible to hear words spoken by a person close to us. These are illustrations of a sound being *masked* by the noise. The amount of masking depends on both the frequency and the loudness of the masked tone. The degree of masking is defined as the required increase in the intensity in dB of the masked sound in order that it can just be heard in the presence of the masking sound. For example suppose we have a pure tone whose frequency is 1000 Hz and whose sound intensity level is 20 dB above auditory threshold. Next we additionally turn on a second tone whose frequency is 400 Hz and whose sound intensity level is 60 dB. We now find that the 1000 Hz tone is not audible until its sound intensity level is raised to 60 dB. The amount by which the 400 Hz tone has masked the 1000 Hz tone is therefore 40 dB. Experiments have shown that (1) the masking of one pure tone by another is most apparent when both tones have about the same frequency and (2) generally the masking tone of frequency f masks tones of frequencies greater than f more effectively than it does of tones of frequencies lower than f. These facts explain why low-frequency masking sounds are generally more effective than high-frequency masking sounds, such as the sounds of jet planes, of trucks, of hums in electronic equipment, etc. An explanation of this is afforded by considering the aural harmonics which are generated by the masking tone. In the numerical example above, the 400 Hz tone has the harmonics 800 Hz, 1200 Hz, 1600 Hz, etc. which have loudnesses comparable to that of the fundamental. Therefore one of these harmonics is near the frequency of 1000 Hz and will effectively mask it. On the other hand, if a 1500 Hz tone at 60 dB were used instead of the 400 Hz tone, the aural harmonics would be 3000 Hz, 6000 Hz, etc. all of which are above the 1000 Hz tone which would not be masked: In a noisy room a person is apt to speak more loudly to overcome the effect of masking and a partially deaf person can hear better than those with no hearing defect. Masking is of significance in musical orchestration. Here the masking of the sounds of some instruments by the sounds of other instruments produces a marked effect on the tonal balance or quality of the sound.

6.4. Quality and Harmonic Content

The third subjective characteristic of a particular sound is known as *quality*. It is also called *timbre* or *tone color*. *Quality* is that characteristic of sound that enables one to distinguish it from other sounds having the same pitch and loudness. It is the characteristic that accounts for the ear's ability to recognize, by hearing only, whether a violin, piano, trumpet, oboe or other instrument is emitting a tone of the same pitch and loudness. We have learned that sound waves are produced by bodies that vibrate and that most sound sources, including the musical instruments, do not produce simple tones of a single frequency, like the tuning fork. Instead, they produce complex tones characterized by complex wave representations that are a combination of simple tones which possess a variety of amplitudes and frequencies. In Figure 2.3 we have shown such a complex wave form containing only a first harmonic and a third harmonic whose amplitude is half that of the first harmonic. We have also seen in Section 2.4 that any complex periodic wave or tone may be represented by a superposition of, or be considered as composed of, a number of simple harmonic waves or tones. The individual simple tones that compose the complex tones are called *partial tones* or just *partials* with the partial of the lowest frequency designated as the fundamental. The partials above the fundamental, called upper partials or overtones, may have frequencies that are integral multiples of the fundamental and in this case the fundamental and upper partials are designated harmonics, as we have witnessed in our study of vibrating strings and resonant air columns. However, in the cases where the frequencies of the partials are not exact multiples of the fundamental they are designated *inharmonic partials*. That which determines to a significant extent the *quality* or *timbre* of any complex tone are the number, distribution, and relative amplitudes or relative intensities of the individual partials that enter in its formation.

Modern electronic wave analyzers are capable of quickly analyzing a complex tone into its composite pure tones, yielding the frequencies of the harmonics present and their relative amplitudes or sound level intensities. To exhibit the results of such a spectrum a plot is made showing the frequencies of the numbers of the harmonics and their relative amplitudes, intensities or sound intensity levels. For example Figure 6.5 shows the spectrum of the wave form of Figure 2.3 with the ordinate expressed as the ratio of the amplitude of the harmonic to that of the most intense partial which here is the fundamental, and the abscissa giving the number of the harmonic. In Figure 6.6 are shown in (a) the complex wave form (dashed curve) whose components are a fundamental, a second harmonic whose amplitude is half that of the fundamental, and a third harmonic whose amplitude is one-third that of the fundamental. These examples illustrate the fact that differences in waveform due to differences in harmonic content result in large part to a difference in quality or timbre. Additional spectra of sounds issuing from tones of the same instrument and from different instruments are indicated in Figure 6.7. The ordinates are expressed as the ratio of the sound intensity level of the harmonic to that of the most intense partial. To each spectrum there is a corresponding wave form, which we have not shown. In (a) and (b) are spectra (up to the 15th harmonic) of sustained tones of the violin open A and open C strings respectively, each exhibiting a characteristic harmonic content and possessing a characteristic different waveform and quality. In (a) the fundamental is the most intense while in (b) the second harmonic bears that distinction. The spectra of some tones of several other instruments are shown in (c), (d) and (e) and of the voice in (f). Each is characterized by a different number and distribution of harmonics with different relative intensities, and hence a different wave form and quality or timbre. Differences in wave form and quality are also a function of the loudness of a tone, the kind of instrument such as a grand piano and a spinet, the manner of attack (growth or onset) of the note, whether the note is a result of plucking, bowing, striking, scraping, or muting (softening the note by a notched clamp made to fit on the bridge of a violin, cello, etc.) and to a very small and essentially insignificant effect of the phasing of its component partials.

Now what is it precisely that enables the ear to discriminate between two different instruments? In the reverse process of sound spectrum analysis it is possible to *synthesize* a complex tone by combining the requisite number of harmonics properly distributed with the proper relative intensities. With modern electronic techniques it is also possible to remove one or more components from a complex tone. In experiments of this kind of synthesis and adjustment two different instruments may be made to emit the same tone with the same harmonic content, such as in the removal of the upper partials of a flute and clarinet, and yet the ear is still able to discriminate between the different instruments. Something about the quality enables the ear to identify the instrument. Researches have led to investigations concerning the presence of some representative sound spectrum unit that would be characteristic of a given instrument or concerning the mode of production of the tone to which the ear displays an instrumental discrimination. Thus there have grown up two theories that attempt to explain what it is to which the ear reacts in discriminating between the sound emitted by one instrument from that of another. One theory stipulates that each different instrument has one or more regions or groupings of harmonics called *formants*, in its tone spectrum, in which the harmonics are relatively prominent and that the number and positions of the formants characterize the instrument and account for the ear's instrumental discrimination. The other theory is based on the fact that the vibrational characteristics of musical tones are *transient* in nature, that is, they have certain build-up and decay times whose particular values contribute significantly to the over-all quality as judged by the ear. In the onset of a tone its growth and loudness with time can vary to a great extent and depend on the manner of attack as with the oboe or the violin. Tones that are recorded without the initial transient are sometimes more difficult to correlate with the instrument. In the complete transient sound envelope there is a short duration growth period, a shorter duration steady state period and a short duration decay period which the theory indicates is not significant; only the growth period is important. *Vibrato*, purposefully introduced in many musical instruments, is a periodic variation of the frequency of a tone resulting in what is called *frequency modulation*. Often this is accompanied by a variation in the amplitude of the tone and termed *amplitude modulation*. This frequency and amplitude modulation accompanying vibrato affect the waveform and have a bearing on the overall quality of the sound. It appears then that the overall quality or timbre is a function of some other factors in addition to a particular harmonic structure and that is why, in our discussion of this characteristic, we indicated that the particular harmonic content and structure determines only to a significant extent, and not solely, the quality or timbre of a complex tone.

Thus far we have been considering only sounds that contain harmonic partials. There are other sounds that consist of a fundamental and partials that are inharmonic such as the sounds from bells, drums, xylophones, tympani and most of the percussive instruments. The beauty of a tone depends on the distribution of its harmonics, and the number of harmonics present determines the richness of the tone. From Figure 5.4 we see that an open and closed pipe sounding the same pitch and intensity have a different quality; the open pipe, containing both the even and the odd harmonics, is much richer in tonal quality. In voice and some instruments inharmonic components are ordinarily present and these may add to tonal richness. However, in most cases the presence of a large number of inharmonic partials causes the tone to lose its musical quality and the sound may become disagreeable, with the sound being termed *noise* as with the bow scratch on a violin. When the sound consists of a combination of all the frequencies in the audible range the spectrum is called *white noise* and this is analagous to white light which consists of a mixture of all visible frequencies or colors. The sound produced by snare drums, cymbals, or escaping air from a constricted valve are in large part white noise.

Problems

6. 1.   In listening to music from a loudspeaker which does not reproduce tones whose frequencies are below 200 Hz, we are able to hear the bass notes of frequencies less than 200 Hz. How do you explain that?

6. 2.   Explain the relationship between and describe the differences of the terms pitch and frequency.

6. 3.   Using Figure 6.3 compare the ranges in frequency of the following instruments: harp and accordian, picolo and flute, oboe and bassoon, violin and bass.

6. 4.   Equation 6.1 states that I is directly proportional to $f^2 A^2$ if ρ and v are constant. Find the ratio of the intensities of two sound waves one of which has three times the frequency and one-half times the amplitude of the other.

6. 5.   Two sound waves have the same amplitude but the frequency of one is 440 Hz while that of the other is 110 Hz. What is the ratio of their intensities?

6. 6.   What are the units for I in Equation 6.1 if the units for ρ, v, f and A are respectively gm/cm^3, cm/sec., Hz and cm.

6. 7.   Show that the units for I in Equation 6.2 are $ergs/sec. \, cm^2$ if p is in $dyne/cm^2$ and v is in cm/sec.

6. 8.   Coupling the variation of the intensity with pressure, as given by Equation 6.2, with the fact that the intensity varies inversely as the square of the distance from a point source of sound what can you conclude regarding the relation between the pressure and the distance?

6. 9.   Resorting to appropriate sources describe qualitatively the operation, function and application to the study of the physical basis of musical sounds of the following instruments: the audio oscillator, the strobscope, the oscilloscope.

6.10.   Relative to the threshold of hearing intensity 10^{-16} watt/cm^2, find the sound intensity level of the following intensities: (a) 2 x 10^{-16} watt/cm^2, (b) 10 x 10^{-16} watt/cm^2, (c) 100 x 10^{-16} watt/cm^2, (d) 1000 x 10^{-16} watt/cm^2, (e) 9 x 10^{-16} watt/cm^2, (f) 8 x 10^{-5} watt/cm^2. Use the table of logarithms in the Appendix where necessary.

6.11.   Relative to the reference pressure level of 2 x 10^{-4} dynes/cm^2 find the sound pressure level of an effective sound pressure of 0.633 dyne/cm^2.

6.12.   Two sounds differ in sound intensity levels by 60 dB. How many times more intense is the louder sound than the fainter sound?

6.13.   What is the difference in sound pressure level between two sounds whose pressures are 25 dynes/cm^2 and 0.025 dyne/cm^2?

6.14.   What change in the loudness level in phons at 300 Hz is produced by a change in the sound intensity level from 20 dB to 45 dB? (Hint: Use Figure 6.4.)

6.15.   What is the change in the sound intensity level at 500 Hz corresponding to a change in loudness level from 30 phons to 60 phons? (Hint: Use Figure 6.4.)

6.16. What is the sound intensity level sum of two sounds each of sound intensity level 70 dB at the same frequency? (Hint: It is not 140 dB).

6.17. What is the sound intensity level of two sounds of sound intensity levels 80 dB and 90 dB at the same frequency?

6.18. What is the usefulness of sound spectra?

6.19. A wave analyzer yields the following sound intensity levels in dB of a violin tone corresponding respectively to the first seven harmonics: 24, 38, 35, 28, 37, 21, 24. Make a spectrum plot of this data in which the ordinate is the relative intensity level and the abscissa is the harmonic number, considering the most intense component as 1.0

6.20. Make a spectrum similar to that in Problem 6.19 but for a voice tone whose sound intensity levels in dB for the first ten harmonics are respectively 44, 45, 26, 7, 14, 21, 20, 10, 12, 7.

CHAPTER 7

Intervals, Scales, and Temperament

7.1. Musical Intervals.

We have seen that the pitch of a tone is associated with its frequency and, even when the tone is complex, the pitch is either determined by the fundamental or is supplied as a subjective tone. The separation or spacing of two tones, in a series of musical tones, is identified with and expressed as the ratio of the frequencies of the tones. Therefore a *musical interval* is defined as *the ratio between the frequencies of any two tones.* For example, the interval between two tones whose fundamental frequencies are in the ratio 2:1 is called an octave and is exemplified by the frequencies 440 Hz and 220 Hz or 660 Hz and 330 Hz. Additional examples are the frequencies 440 Hz and 330 Hz whose ratio reduced to lowest terms is 4/3 and known as the interval of a fourth, or again the frequencies 396 Hz and 264 Hz whose ratio is 3/2 and known as the interval of a fifth.

In Chapter 5 it was shown that the vibrations of strings and resonant air columns give rise to partials which are harmonics whose frequencies are integral multiples of the fundamental. The tones of the fundamental and its harmonics thus form the harmonic series. In Figure 7.1 are shown on the musical staff the first twelve partials of the harmonic series based on the letter designation C_2 (see Figure 6.3) as the fundamental, with the indicated frequencies taken from the just scale which we discuss in Sections 7.3 and 7.4. For the tones in parenthesis, Bb and C#, the frequencies are approximate.

It is often necessary to deal with the sum of two or more intervals. For example the interval between G_3 and E_4 in Figure 7.1. Now the interval between G_3 and C_4 is 264/198 or 4/3 and that between C_4 and E_4 is 330/264 or 5/4. Since the ratio of the frequencies for G_3 and E_4 is 330/198 or 5/3 we see that the interval between G_3 and E_4 is given by the product of the interval between G_3 and C_4 and the interval between C_4 and E_4 or $(4/3)$ x $(5/4)$ = $5/3$. This is an example of the general rule that *the sum of two or more intervals is equal to the product of the frequency ratios that represent the intervals.* The student may wish to verify this by the example of multiplying the frequency ratios of the successive intervals from C_4 to C_5 and obtain the interval for the octave.

A musical name of an interval such as a third, a fourth, a fifth, an octave, etc. is made in accordance with the following rule: The numerical name of the interval is determined by counting inclusively the consecutive alphabetic designation from the lower note to the higher note. Thus the octave from C_3 to C_4 contains the eight notes C_3, D_3, E_3, F_3, G_3, A_3, B_3, C_4, which also involves eight lines and spaces on the musical staff (or only seven steps). This designation is made irrespective of any accidentals such as sharps, flats or naturals. The numerical name and corresponding frequency ratio of the intervals in common use are as follows: unison, $1/1$; major third, $5/4$; minor third, $6/5$; fourth $4/3$; fifth, $3/2$; major sixth, $5/3$; minor sixth, $8/5$; major seventh, $15/8$; minor seventh, $9/5$.

7.2. Consonance and Dissonance: Musical Scales

In the audible range there are an infinite number of frequencies but for musical purposes only a finite number has been chosen and these frequencies for the most part involve intervals that are pleasing to the human ear. When two or more tones are sounded together or in rapid succession certain combinations or intervals produce more agreeable auditory effects than others. Even in the rendition of a melody by an instrument or the voice, in which there is emitted a succession of single tones, there occurs reverberation in a confined space or the persistence of sound, analogous to the persistence of vision, and certain successive musical intervals, which thus overlap, produce smoother sensations than others. When two or more tones, sounded together or in succession, produce a pleasing or smooth auditory sensation we say the sound is *consonant*; when they produce an irritating, harsh or rough sensation we say the sound is *dissonant*. Hemholtz attributed the disagreeable dissonant effects in Western music to the presence of disturbing beat frequencies either between the fundamental tones themselves or between some of their harmonics. He indicated that a beat rate between about 30 and 130 Hz had the effect of producing an aural irritation with the maximum effect around 33 Hz. Thus in the mid-range of the piano keyboard (see Figure 6.3), the beat rate between adjacent notes whose interval is a semitone, as between B_4, 494 Hz, and C_5, 523 Hz, is about 29 Hz, and between two adjacent whole tones as between C_5 and D_5, 587 Hz, is about 64 Hz. These beat rates are between the fundamental frequencies. If we consider the tone B_1, 62 Hz, its second and third harmonics are 124 Hz and 186 Hz. The fundamental, second and third harmonics of D_2, 73 Hz, are respectively 146 Hz and 219 Hz. When B_1 and D_2 are sounded together the beat rate between the fundamentals is 11 per second which is not in the disturbing range while that between the third harmonics is 33 per second which results in pronounced harshness. Computations of this kind will show that intervals that give rise to dissonant beat rates in the lower register of the keyboard produce consonant beat rates in the upper register. It is the pulsing of the sound due to the beats that causes auditory roughness analogous to the disturbance of flickering light on the sensation of sight. However musical fashions and tastes are not only different in countries with different cultures but also change with time. Combinations of tones that were in the past considered dissonant in Western music are today tolerated by the ear and even considered consonant by many. For ears accustomed to the classical works of Bach, Beethoven, Haydn, and Mozart much of the music written in modern times is highly dissonant but apparently consonant to ears of much of the present generation. What is termed a dissonance or a consonance by a person involves the person's aesthetic and psychological background.

It was early recognized in the Western part of the world, and is generally agreed upon, that certain tones when sounded together produced a pleasing auditory sensation and that the most pleasing combination of two tones is that in which the interval or frequency ratio can be expressed by two small integers such as the ratio 2:1 (octave), 3:2 (fifth), 4:3 (fourth), 5:4 (major third), 5:3 (major sixth), and 8:5 (minor sixth). These are intervals which, Figure 7.1 reveals, are found in the harmonic series. From such pleasing combination it is possible to build up a *musical scale* which is *a series of musical tones increasing or decreasing in frequency by definite musical intervals suitable for musical purposes.* If the scale (the word scale is derived from the Latin word *scala* meaning steps of a ladder) contains only intervals that are present in the harmonic series it is known as a *just diatonic* scale or a scale of *just intonation*. We shall see that the scales of just intonation possess significant disadvantages and practicable shortcomings which resulted in its having been replaced by the scale of *equal* or *even temperament*. We shall not go into a detailed historical review of the evolution of musical scales including those developed by different cultures but shall concentrate on the just diatonic and equal temperament scales in the sections that follow in this chapter.

7.3. The Just Major Diatonic Scale

We have indicated that the interval of a just major third, 4:5, produced a pleasing auditory effect. If to this major third is added a just minor third, 5:6, there results the pleasing three-tone combination, called a *major tirad*, which has been the foundation of Western music for several centuries and whose relative frequency ratios are 4:5:6. The just major diatonic scale contains eight tones to the octave and is composed of three sets of major triads structured as follows:

C	D	E	F	G	A	B	C'	D'
4		5		6				
			4		5		6	
				4		5		6

The eight notes are represented by the letters C to C' where the starting note C is designated the *tonic* or keynote and C' has a frequency that is double that of C. When the tonic or keynote is changed the three triad ratios remain in the indicated positions and the notes of the scale are related to the new tonic as starting point. Letting f_c be the frequency of the tonic we can obtain the frequencies of the remaining tones of the scale f_D, f_E, etc. in terms of f_C by simple proportion as follows:

$f_E/f_C = 5/4$; $f_{/E} = (5/4)f_C$

$f_G/f_C = 6/4 = 3/2$; $f_G = (3/2)f_C$

$f_F/f_{C'} = 4/6 = 2/3$; $f_F = (2/3)(2f_C) = [4/3]f_C$

$f_B/f_G = 5/4$; $f_B = (5/4)f_G = (5/4)[(3/2f_C)] = (15/8)f_C$

$f_A/f_{C'} = 5/6$; $f_A = (5/6)(2f_C) = (5/3)f_C$

$f_{D'}/f_G = 6/4 = 3/2$; $2f_D = (3/2)f_C = (3/2)[(3/2)f_C]$;

$f_D = (9/8)f_C$

Aligning these major diatonic frequencies in ascending order corresponding to the scale letter designations we can find the intervals or frequency ratios between successive tones as follows:

Table 7.1. Frequencies in terms of the tonic and frequency ratios of adjacent tones in the major scale of just intonation.

Name of Note	C	D	E	F	G	A	B	C'
Frequency in terms of tonic f_c	f_c	$9/8f_c$	$5/4f_c$	$4/3f_c$	$3/2f_c$	$5/3f_c$	$15/8f_c$	$2f_c$
Intervals or frequency ratios between adjacent tones		$9/8$	$10/9$	$16/15$	$9/8$	$10/9$	$9/8$	$16/15$

The ratios in the third row of Table 7.1 are obtained by dividing the frequency of any note by that of the preceding note. For example the frequency ratio between the notes F and E is $(4/3)f_c \div (5/4)f_c = 4/3 \times 4/5 = 16/15$. We notice that the just major diatonic scale is composed of only three intervals between adjacent notes, namely, $9/8$, $10/9$ and $16/15$. The ratio $9/8 = 1.125$ is termed a *major whole tone*, the ratio $10/9 = 1.111$ is termed a *minor whole tone*, and the ratio $16/15 = 1.067$ is termed a *major* or *diatonic semitone*. Note that two semitone steps are greater than a major whole tone step. It is clear that if the frequency of the tonic or of any other note in the scale is known or assigned, the frequencies of the remaining members of the scale can readily be obtained by simple multiplication or division. If the frequency of the tonic is given, it is best to use the relationships in the second row to obtain the frequencies of the additional scale notes for then an error in the computation of one note will not be carried along in the notes that follow as would occur if the ratios in the third row were used. If it is desired to change the tonic, say the scale in the key of D, we start with D and obtain E by multiplying the frequency of D by $9/8$, then obtain F by multiplying the frequency of D by $5/4$, etc. The Greeks have used many scales including the major diatonic scale and the minor diatonic scale which we take up in the next section.

7.4. The Just Minor Diatonic Scale

The just minor diatonic scale is composed of three sets of minor triads whose fundamental frequencies satisfy the ratios 10:12:15 as shown in the following array using again C as the tonic:

C	D	E	F	G	A	B	C'	D'
10		12		15				
			10		12		15	
				10		12		15

Proceeding as before to obtain the frequencies of the notes in the scale in terms of the frequency of the tonic f_C we have

$$f_E/f_C = 12/10; \quad f_E = (6/5)f_C$$
$$f_G/f_C = 15/10; \quad f_G = (3/2)f_C$$
$$f_F/f_{C'} = 10/15; \quad f_F = (2/3)(2f_C) = (4/3)f_C$$
$$f_A/f_{C'} = 12/15; \quad f_A = (4/5)(2f_C) = (8/5)f_C$$
$$f_B/f_G = 12/10; \quad f_B = (6/5)(3/2)f_C = (9/5)f_C$$
$$f_{D'}/f_G = 15/10; \quad f_D = (1/2)(3/2)[(3/2)f_C] = (9/8)f_C$$

Again proceeding as before we obtain Table 7.2 which lists the frequencies of the tones in terms of the tonic frequency and the frequency ratios of the adjacent tones in the just minor diatonic scale.

Table 7.2. Frequencies in terms of the tonic and frequency ratios of adjacent tones in the minor scale of just intonation.

Name of Note	C	D	E	F	G	A	B	C'
Frequency in terms of tonic f_C	f_C	$9/8f_C$	$6/5f_C$	$4/3f_C$	$3/2f_C$	$8/5f_C$	$9/5f_C$	$2f_C$
Intervals or frequency ratio between adjacent tones		9/8	16/15	10/9	9/8	16/15	9/8	10/9

Inspection of the third row of Figure 7.2 reveals that the just minor diatonic scale is composed of the same three intervals 9/8, 10/9 and 16/15 as are present in the just major diatonic scale but they occur in a different order.

The diatonic scales, with fundamental intervals of the harmonic series, evolved, as we know it today, in Western music. In primitive and different cultures aesthetic and practical considerations, more than physical laws, governed the scale development. Diatonic and modified diatonic scales are to be found in ancient times and in different cultures. The Egyptians, the Hindus, the Chinese, the Persians and the Greeks pursued diverse lines of scale developments and there arose the *pentatonic, hexatonic,* and *heptatonic* scales with five tones, six tones and seven tones respectively to the octave. The Arabs and Persians divide their octave into a large number of intervals and the quarter-tone interval is frequently employed, making the sound rather strange to the Western ear. Certain oriental scales, as well as primitive scales of certain races, contain intervals that for the most part do not resemble those of the diatonic scale except for the intervals of the octave and the fifth which appear to be common.

Pythagoras, the Greek philosopher and mathematician of the 5th Century B.C., devised a scale that bears his name and to which the diatonic scale appears to be related. The scale is built up in twelve cyclic steps upward from the tonic using only two intervals; the octave 2:1 and the fifth 3:2. These twelve steps spanning an octave are produced by twelve steps of consecutive fifths upward. Each step involves a multiplication by 3/2, and when such an operation results in a tone that exceeds the compass of one octave it is brought back into the octave range by transposition which involves multiplying by 1/2. Upon application of the twelfth step the resulting tone, instead of being ideally an octave higher than the tonic, and hence essentially returning to a starting tone, comes out somewhat sharp as represented by the interval 531441/524288, or 24 cents (see Section 7.6), which is known as the *Comma of Pythagoras.** The fact that the pythagorean system never returns on an octave note can be seen simply from the observation that in advancing upward by fifths and the necessity of bringing a tone into the compass of an octave always results in an interval of $3^a/2^b$ where a and b are integers. Since 3^a can never equal 2^b this ratio can never be unity. The pythagorean comma, along with the fact that the pythagorean scale contains two different sizes of

*For more detailed development of the pythagorean scale see J.M. Barbour and F.A. Kuttner, "Meantone Temperament in Theory and Practice," Musurgia Records, New York, 1958.

semitones which is very awkward, are impurities and shortcomings of
the system which many workers attempted to overcome by various
adjustments and compromises called temperaments.

7.5. Analysis of the Just Diatonic Scale or System

To evaluate the practicability of the just diatonic scale we shall first
compute a set of frequencies in the key of C major using the intervals
in Table 7.1, and then examine the frequencies obtained when we
transpose to the scale of D major and the scale of E major. The
computed frequencies in hertz are shown in Table 7.3, and the
numbers have been rounded off to the next higher integer where the
decimal number is 5 or greater. In the key of C major the tonic was

Table 7.3
Comparison of frequencies in transposed major diatonic keys

Name of Note	Key		
	C major	D major	E major
C	264		
D	297	297	
E	330	334*	330
F	352	371*	371*
G	396	396	413*
A	440	446*	440
B	495	495	495
2C	528	557*	550*
2D	594	594	619*
2E	660	668*	660

adjusted to yield 440 Hz for A and the frequencies for D,E,F,...etc.
were obtained by multiplying 264 Hz by the respective interval ratios
9/8, 5/4, 4/3 ...etc. or their decimal equivalents 1.125, 1.250, 1.333 ...
etc. respectively. The tonic for the key of D major, 297 Hz, is the value
for D appearing in the C major scale and the frequencies in this
column are obtained by multiplying 297 Hz by the same interval
values 9/8 to yield E, 5/4 to yield F, 4/3 to yield G, and so on. A similar
procedure was followed to obtain the frequencies of the E major scale.
The notes marked with an asterisk * are new notes that must be added

to those of the C major scale to permit the desired transposition or modulation. For these two major scales eight new notes are required and additional ones would be needed to permit transposition to the other major and minor just diatonic keys. Again when we compare the frequencies of the C major scale with those of the C minor scale using the same tonic for each, we observe in Table 7.4 that three new notes need to be added to the C major scale to permit transposition to the C minor scale. When this kind of computation is carried out for the 15 signatures it would be necessary to provide about 72 notes to the octave

Table 7.4
Comparison of the frequencies in the C major and C minor diatonic scales

Name of Note	Key	
	C major	C minor
C	264	264
D	297	297
E	330	317*
F	352	352
G	396	396
A	440	422*
B	495	475*
2C	528	528

in order to permit free transposition or modulation for all possible changes of key. Some of the compiled frequencies might not have to be added since their deviations from the diatonic frequencies are small enough that they may be unnoticeable by most listeners. However the remaining number of tones per octave required is still of such magnitude as to make it impracticable for instruments having fixed tones such as the piano, the organ, the flute and most of the wind instruments.

Although the just diatonic scale does not have all of its intervals that may be categorized as producing a pleasing auditory sensation it possesses a rather large number of consonant intervals. In the search for a scale that adheres as closely as possible to the just diatonic scale but is practicable for fixed keyboard instruments and permits free transposition or modulation various schemes have been developed, each with its advantages and disadvantages. The compromise system that is practicable, permits free modulation, and has no interval in any key that sounds too much out of tune, is known as *equal* or *even temperament* which has been in vogue in Western music for about two centuries. Before we discuss this equally tempered scale we shall first devote the next section to a description of a convenient and desirable means of expressing intervals in terms of fractions of a semitone and designated by a unit called the cent.

7.6. Cents

In the *cent* method of calculating musical intervals the octave is considered divided into 1200 equal parts each interval of which is called one cent and the interval of an octave is equal to 1200 cents. Since for an octave the sum of the 1200 equal intervals must add to the interval 2/1 and since in adding intervals we multiply their frequency ratios then, if we designate the interval of a cent by the symbol ¢, we have

$$1¢ = \sqrt[1200]{2} = 2^{1/1200}$$

and for any number of cents N we have

$$¢^N = 2^{N/1200} = r \text{ or } 2^{N/1200} = r \tag{7.1}$$

where r is the frequency ratio or interval corresponding to the sum of N cents. Equation 7.1 relates any interval of frequency ratio r to the equivalent number of cents N. We can obtain a more convenient form for the computation of N by taking the logarithm of each side of Equation 7.1,

$$\frac{N}{1200} \log 2 = \log r \qquad\qquad N = \frac{1200}{\log 2} \log r$$

and using the value 0.30103 for the logarithm of 2 on the base 10 we obtain

$$N = 3986 \log_{10} r \tag{7.2}$$

where $\log_{10}$ indicates the logarithm on the base 10. Equation 7.2 is the convenient expression that will readily yield the number of cents corresponding to any musical interval r.

For example, for a perfect just diatonic fifth where r = 3/2 Equation 7.2 yields

$$N = 3986 \; \log_{10} \; 3/2 = 3986 \; \log_{10} \; 1.5 = 3986 \; (0.176) = 702 \text{ cents}$$

As another example, for the diatonic major whole tone 9/8, r = 1.125 and $N = 3986 \log_{10} 1.125 = 3986 (0.05115) = 204$ cents. We see that the method of cents designation is essentially a logarithmic computation in terms of musical intervals and therein lies its advantage. Whereas to obtain the sum of two or more intervals we multiply their frequency ratios, in the cents computation we simply add the cents of each interval since the logarithm of the product of two factors (intervals) is the sum of the logarithms of the individual factors.

7.7. The Equally Tempered Scale

The system of equal or even temperament is credited to Johann Sebastian Bach who strongly advocated its adoption early in the eighteenth century. In this system the octave is divided into twelve equal intervals or equally tempered semitones. Since there are 12 equal intervals each of size or value x their product x^{12} expresses the interval sum for the octave or

$$x^{12} = 2$$

and
$$x = \sqrt[12]{2} = 1.05946 \tag{7.3}$$

This then is the value of the semitone interval on which basis the frequencies of the tones on the piano are obtained as indicated in Figure 6.3. This is to be compared with the diatonic semitone 16/15 = 1.06667 the difference being about 0.7 percent or about 12 cents. Since there are 12 equal steps in the octave each tempered semitone is 100 cents or the cent is 1/100 of a tempered semitone. In terms of cents the tempered fifth, being seven semitones is 700 cents; the tempered third, being four semitones, is 400 cents; etc.

Let us now see how the vibration numbers or frequencies of the equally tempered scale compare with those of the just diatonic scale. Table 7.5 shows this comparison with the A major diatonic scale where we have chosen the tonic A of frequency 220 Hz and used the name of the note as that of the equally tempered piano. The equally tempered frequencies are obtained by using the semitone interval 1.05946 as a successive multiplier and the difference value is the equally tempered value minus the just diatonic value. We note that only the tonic and its octave are in agreement and, with respect to the

Table 7.5.
Comparison of the equally tempered and just
diatonic scales based on A 220 Hz as the tonic

Name of Note

	A	B	C#	D	E	F#	G#	A
Diatonic Scale	220	247.50	275.00	293.33	330.00	366.66	412.50	440
Equally Tempered Scale	220	246.94	277.18	293.66	329.63	369.99	415.30	440
Difference	0	-0.56	+2.18	+0.33	-0.37	3.33	2.80	0

just diatonic scale, C#, D, F# and G# are too sharp while B and E are too flat. However, B, D, and E are out of tune only slightly and would be scarcely discernable when two comparison frequencies are sounded together. On the other hand C#, F# and G# are out of tune by a discernable and objectionable amount, again when the two comparison frequencies are sounded together. This is illustrative of the general observation that in the equally tempered scale the fourths and fifths, which are predominantly significant in melody and harmony, are very nearly in tune, when compared with the corresponding diatonic intervals, while the major and minor thirds, sixths and sevenths are somewhat out of tune. Table 7.6 indicates these facts as well as additional ones. Column two lists the frequency ratios of the notes of the major diatonic scale in relation to the tonic C taken as 1.000, column three lists the frequency ratios of the notes of the equally tempered scale relative to the tonic C, column four lists the cents equivalent of the values in column two, and column five lists the cents equivalent of the values in column three or of the equally tempered semitones. Comparison of columns two and three reveals in general for any octave range the same features illustrated by our discussion of Table 7.5. These features may best be seen by comparison of columns four and five. Thus the diatonic fifth is 702 cents while the tempered fifth is 700 cents, giving a melodically negligible difference of 2 cents or 0.02 of a semitone. Likewise for the fourths the difference is only 1 cent or 0.01 of a semitone. For the thirds, sixths and sevenths the differences are respectively 400-386 or 14 cents, 900-884 or 16 cents and 1100-1088 or 12 cents, all of which we have already indicated are not melodically negligible. The comparison with the minor diatonic frequencies reveals similar results. In the first column are listed the names of the notes, showing those appearing in the diatonic scales and

Table 7.6
Frequency ratios in the scale of equal temperament and comparison with those of the diatonic scale; both based on C as a tonic

Name of Note	Major Diatonic Frequency Ratio	Equally Tempered Frequency Ratio	Number of Cents of the Diatonic Intervals	Number of Cents of the Tempered Intervals
C	1.000	1.000	0	0
C# or Db		1.059		100
D	1.125	1.122	204	200
D# or Eb		1.189		300
E	1.250	1.260	386	400
F	1.333	1.335	499	500
F# or Gb		1.414		600
G	1.500	1.498	702	700
G# or Ab		1.587		800
A	1.667	1.682	884	900
A# or Bb		1.782		1000
B	1.875	1.888	1088	1100
C	2.000	2.000	1200	1200

the five sharp or flat notes that are added in the equally tempered scale corresponding to the black keys of the piano. In the diatonic scale and other earlier scales *enharmonic* notes such as C# and Db do not have the same frequency but in equal temperament they are identical. The sharp and flat pairs, in column one of Table 7.6, that differ in notation but have the same frequency are said to have an *enharmonic* relationship. Since the semitones are the same in the equally tempered scale, all the 12 intervals of an octave are the same, and the ratio of any value in column three to the note preceding it is the semitone value 1.059. We may therefore compute the frequencies of any note in any scale by using the values in column three.

This compromise system, although lacking in true intonation, since only the octave is perfectly in tune, permits of modulation or transposition into any of the fifteen signatures and utilizes only twelve notes per octave. The practicable advantage of the equally tempered scale over the just diatonic scale have outweighed its disadvantages and it has stood the test of time in remaining the accepted basic system of musical development in the Western world.

Modern electronic instruments provide a visual means of indicating when a string has been adjusted in tension to emit a tone of a fundamental frequency and are thus excellent for use in the accurate tuning of the piano in equal temperament. Although this method of tuning is being used more widely there still are many tuners who resort to the tuning of the piano by the auditory beat method. In this method one octave of the temperament is laid usually in the middle of the keyboard where the standard pitch of A 440 Hz is set and included. This is done by tuning each string by means of auditory beats between certain upper partials of a series of pairs of notes usually fifths and fourths. For those notes that are emitted by the key hammer simultaneously striking more than one string, only one string is permitted to vibrate by the proper insertion of rubber or felt wedges to damp the other strings of the note to comparative silence. Once the temperament has been laid, the wedges are removed and the unisons are tuned to yield no beats. Finally the remaining tones are set by tuning octaves, also with no beats, above and below the middle octave. For a more detailed account of the tuning procedure the student is referred to the following references: John Redfield, "Music: A Science and an Art," Alford A. Knopf, New York, 1927: Charles A. Culver, "Musical Acoustics," McGraw-Hill Book Co., Inc., 1956.

Problems

7. 1.   Find the musical interval between the following pairs of frequencies and identify them by their numerical names: 264 and 330, 297 and 495, 330 and 495, 440 and 528.

7. 2.   Find the sum of the following intervals and identify the result with its numerical name: 16/15, 10/9 and 9/8.

7. 3.   Using Helmholtz's contention that dissonance is due to the disagreeable sensation produced by beats would there be dissonance when sounding simultaneously the tones C_2 whose frequency is 65 Hz and E_2 whose frequency is 82 Hz?

7. 4.   (a) What are the frequencies of a perfect (diatonic) fifth above and a perfect (diatonic) fourth below 440 Hz?   (b) How are these frequencies related?

7. 5.   (a) What are the frequencies of a perfect (diatonic) fourth above and a perfect (diatonic) minor third below 330 Hz?   (b) What is the interval of these frequencies?

7. 6.   Verify that the following musical intervals are found in the harmonic series: 2/1, 3/2, 4/3, 5/4, 5/3, 6/5, 8/5.

7. 7.   On the basis of the tonic F with a frequency of 352 Hz compute the frequencies of the just major diatonic scale, the just minor diatonic scale and, by comparison between the two, indicate what notes have different frequencies on the two scales.

7. 8.   Given that the frequency of B on the just major diatonic scale, whose tonic is G, is 495 Hz find the frequencies of D and F.

7. 9.   List the advantages and disadvantages of the just diatonic scale and of the equally tempered scale.

7.10.   Find the cents values of the intervals in Problem 7.6. Use the table of logarithms given in the Appendix.

7.11.   Compute the cents value of the Pythagorean Comma. First divide the fraction for the Pythagorean Comma and then proceed with the application of Equation 7.2.

7.12.   The Pythagorean scale system contains two semitones; the diatonic semitone and the chromatic semitone whose intervals are respectively 256/243 and 2187/2048. Find their values in cents.

7.13.   (a) Find the value in cents of the equally tempered semitone. (b) Is this nearer to the just diatonic semitone or is the Pythagorean chromatic semitone nearer?

7.14.   Compute the frequencies of the equally tempered scale from the tone A, 220 Hz to the tone A, 440 Hz. Use 1.05946 as the multiplier and retain six significant figures for each frequency.

7.15.   It has often been said that the equally tempered scale is the result of having tampered with the scale. Can you explain why?

7.16.   For laying the temperament on the piano find the significant

number of beats per minute between $C_4\#$, 277.18 Hz, and the
fifth above, $G_4\#$, 415.30 Hz, for proper tuning. (Hint: Use the
third upper partial of $C_4\#$ and the second upper partial of $G_4\#$.)

7.17.   Resort to other sources and write a detailed procedure for
tuning the piano in equal temperament by the method of
auditory beats.

CHAPTER 8

Musical Instruments

A musical instrument consists of one or more resonating components that are capable of producing musical tones when actuated or excited. In our treatment thus far we have examined the laws, principles, and important effects that govern the vibration, emission, and characteristics of two significant categories of musical instruments; stringed instruments and the instruments that utilize air columns as their vibrating media. We have also dealt with some sound characteristics of percussion instruments. In this chapter we shall categorize the various families of musical instruments and describe some of those aspects that are particularly significant in affecting the characteristics of the musical sound. It is not our intent to indulge in a detailed description of the construction and make of each instrument; this has been adequately done and may be found in various sources in musical libraries.

8.1. The Family of Stringed Instruments

In Chapter 5 we have shown that a string under tension produces a musical tone that is rich in quality and contains upper partials that are both the even and odd harmonics of the fundamental. We have also indicated that the particular quality is a function of the manner in which the string is excited, that is, whether it is bowed, plucked, or struck.

In the list of stringed instruments that are bowed are the following: violin, viola, violon cello, double base. As representative of this group we consider in some detail the violin which has the smallest physical size but the same general shape and method of construction as the other members of the group. It is provided with four strings stretched between the tailpiece, that is held under the chin of a performer, and the end of the finger board after passing over a bridge supported on the front, also called the belly, of the violin body. The strings are tuned a fifth apart at G_3, D_4, A_4, E_5. The back of the body is coupled to the belly by means of a short wooden post situated approximately under the leg of the bridge on the E string side. The front and back of the body are made of wood and separated by being glued to additional strips of wood called ribs. Under the other leg of the bridge and attached to the underside of the belly on the G string side is a long wooden strip called a bass bar that runs longitudinally; the bass bar helps to strengthen the belly and aids in transmitting some of the vibrations from the bridge to the belly. The belly also contains two S-shaped sound holes that are situated symmetrically on either side of the bridge. The belly, the back, the cavity of the body and the sound holes form a system that is highly resonant and functions to couple efficiently the vibration of the strings to the air. It will be recalled that a vibrating string by itself is not a very efficient emitter of sound and requires a means of coupling to a large radiating surface to increase the sound volume. The overall characteristic of sound quality of the violin is in part due to the air resonance inside the violin box and the wood resonances due to the vibration and radiation from the front and back plates of the body.

The strings, whose tension and proper fundamental frequencies are adjusted at the neck (scroll end) of the instrument by the pegs around which the strings are wound, are set into vibration by drawing a *bow* across them. The bow consists of strands of horsehair stretched between the two ends of a thin shaft of wood and attached at one end to a movable piece called the *frog* which, by a screw arrangement, makes it possible to vary the tension of the horsehair. Ordinarily the bowing is made approximately midway between the bridge and the end of the finger board after the horsehair is rubbed with rosin to increase the friction between it and the strings. When the bow is moved across a string it drags the string with it until the restoring force increases sufficiently to overcome the frictional force and the string springs back under the bow to its normal position. Due to the strings inertia it somewhat overshoots its starting position but the bow again engages it and the cycle is repeated. The driving force possesses a saw-tooth wave shape which contains a fundamental and the even and odd harmonics and thus corresponds to the resonant frequencies of the string which we have seen is rich in quality. Increasing the pressure of the bow increases the intensity of the upper partials and the closer the bow is to the bridge the more prominent the higher partials. Also, the loudness of a tone depends upon the speed of bowing and upon the number of horsehairs in contact with the string and this number is adjusted by varying the angle that the plane of the bow makes with the string. Since the four strings of the violin are capable of producing a large variety of tonal qualities, the violin section of an orchestra is basic in orchestral instrumentation.

The actions of the viola, violon cello, and double bass are the same as that of the violin. The viola, violon cello and double bass have their four open strings tuned as follows: viola—C_3, G_3, D_4 and A_4; violon cello—C_2, G_2, D_3 and A_3; double bass—E_1, A_1, D_2 and G_2. Normally the overall lengths of the instrument and bow for the violin, viola, violon cello and double bass are respectively 23½ inches and 29½ inches, 26 inches and 29½ inches, 46 inches and 28 inches, 78 inches and 26 inches. The strings of this family of instruments may also be plucked, the excitation then being termed *pizzicato*, and this produces tone qualities that are very much different than when the strings are made to vibrate by bowing. The point at which a string is plucked has a bearing on the resulting tone quality (see Section 5.2).

In the list of stringed instruments that are plucked are the following: harp, guitar, mandolin, banjo, ukelele, harpsichord, zither, lyre, lute. Each of these instruments contains a sound board to couple the vibrations of the strings to the air. In general, as large a soundboard as possible is employed since the amount of acoustical radiation from a sound board radiator increases with its size. However in the case of the harp, where vertical strings are supported on a triangular-shaped frame, the lower ends of the strings are connected to a relatively small soundboard. Because of this the strings are not damped to a high degree and consequently the sound persists for a long time.

The principal example of the struck-string instrument is the piano which covers the widest frequency range of 27.5 Hz to 4186 Hz or over seven octaves, and with which we are all familiar. In the method of excitation the string or group of strings of one note is struck at a fixed point, about one-seventh to one-ninth the string length, by a hammer which is felt-covered and set into motion by manually striking or depressing one of 88 keys. A rather complicated mechanism known as the *action* insures that the hammer after striking does not remain in contact with a string and rebounds immediately after striking, that the hammer remains about one-half inch from the string while the key is held down, that the mechanism resets itself when the key has been released less than half way so as to provide a rapid repetition of striking the key, and that the tone persists during the time the key is depressed and stops when it is released. The tone is stopped upon release of the key by the contact of a felt damper with the string. If the musician desires, he may raise the dampers off all the strings by means of the right foot pedal and thus permit the sound of a note or chord to persist while playing other notes or he may shift the entire keyboard mechanism by a second left foot pedal so the hammers strike only two instead of the three strings of a note to obtain a sound of reduced volume and different quality. On grand pianos there is a third pedal that permits, when depressed, only the tones that are played to be sustained and all the other strings are damped normally when their keys are released. The sounding board to which the strings are attached is large and massive, and possesses a broad resonance characteristic so as to reinforce the partials in the piano tones. In accordance with our treatment of quality or timbre it is clear that the different tones of a piano possess different qualities. In addition to the fundamental the spectra of the lower register tones contain a larger number of upper partials than those of the middle and higher registers. Piano strings, being made of steel, possess a large amount of stiffness which affects the restoring force and results in the higher partials of a tone not being exact integral multiples of the fundamental frequency. This departure from a harmonic relationship, which is more pronounced the higher the partial, is called *inharmonicity*. A small amount of inharmonicity appears to add warmth to the tones but when the degree of inharmonicity becomes excessive it detracts from the tonal quality of the instrument.

8.2. The Family of Wind Instruments

In the wind instruments the basic acoustical radiator is the vibrating air column. In Chapter 5 we have studied the resonant air column inside open and closed tubes and learned that there are characteristic discrete modes of vibration resulting in the emission of both the even and odd harmonics by the open tube and the odd harmonics by the closed tube. In a wind instrument there is generated a longitudinal resonant standing wave in its hollow interior and the resulting sound is radiated through openings in the instrument's walls or through the open end of the instrument.

The condensations and rarefactions needed to establish a standing wave in the air columns of musical woodwind instruments are generated in two ways: by an oscillating air stream also known as an *air-reed* and by the vibrations of a mechanical thin section of elastic material called a *mechanical-reed*. All woodwind instruments may be categorized as being of the air-reed type, the mechanical reed type, or the type involving a combination of the two.

Air-reed Instruments

In an air-reed instrument, such as the flue organ pipe shown in Figure 8.1, air that is blown in at the inlet is directed into the mouth of the pipe as a thin sheet thru the channel or flue. When the stream of air is directed onto the lip or sharp edge of the pipe there are formed eddies or vortices on both sides of the edge. As a consequence the stream vibrates back and forth across the edge and generates a sound called an *edge tone* whose frequency depends upon the speed of the air stream. If the frequency of formation of the eddies is near the natural frequency of vibration of the air column in the pipe, resonance occurs with a consequent emission of a loud sound. The air column and the eddies actually form an acoustically coupled system and when the air column is vibrating it has the effect of stabilizing the frequency of eddy formation to that of the vibrating air column. The even harmonics for the open pipe in Figure 8.1 are also produced because it resonates at these frequencies as well as at the fundamental frequency and so controls the oscillating air stream at these frequencies. Flue organ pipes are of the open and closed types. Some air-reed instruments are equipped with holes in the walls and this permits changing the length of pipe and hence the resonant frequency by closing them with the fingers or by means of keys or valves. The air columns are not all of the simple cylindrical or conical shape and such differences play a significant part in the characteristics and tonal quality of the emitted sound. In the flute there are two cylindrical sections that are joined by a conical section. One end of the flute is open and the other end is closed. The *embouchure*, or blowhole, is located near the closed end and the effective length of the resonating air column is regulated by a number of holes which may be closed by the fingers or by keys. The flute performs as an open resonance tube. The edge tone phenomenon occurs whenever there is relative motion between air and a solid body such as the "swinging of a wire" when a wind blowing onto a wire forms eddies at a rate that corresponds to the natural frequency of the wire, or when one blows across an empty or partially filled bottle with the result that the stream of air from the lips impinges on the edge of the bottle opening thus forming the edge tone. This kind of vibratile element that generates standing wave patterns in certain musical wind instruments is aptly named an air-reed. In addition to the flue organ pipe the following wind instruments operate as air-reed instruments: whistle, flute, piccolo, fife, ocarina, recorder and flageolet.

Mechanical-reed Instruments

In the mechanical-reed instrument there is a vibrating mechanical reed, made of an elastic material such as cane or bamboo, which is actuated in a mouthpiece by blowing. This results in intermittent puffs of air that generate the standing waves in the attached air column which vibrates in its various modes. Mechanical-reed instruments are of the single-reed type and the double-reed type.

In the single-reed instruments there are the clarinet, the saxophone, the free-reed organ (the sound output of the reed radiates directly into the air), the reed organ pipe, the accordion, the harmonica and the bagpipe. In playing a reed instrument the standing wave that is generated in the air column contains a velocity loop or pressure node at the first open hole. The oscillations are maintained by blowing and most of the sound emerges through the side holes. In general the reed and the pipe are arranged to have approximately the same frequency and the instrument is tuned by adjusting the length of the vibratile part of the reed thus changing its natural vibrational frequency. In the free-reed instruments, such as the free-reed organ, the accordion, and the harmonica, the sound output of the reed radiates directly into the air. In other mechanical reed instruments the reed is coupled to the resonant air column as in the clarinet, the saxophone, the bagpipe and the reed organ pipe.

In the double-reed instruments there are the oboe, the bassoon, the English horn, and the sarrusophone. In these instruments there are two mechanical reeds that throttle a stream of air and thus produce the musical wave in the air column and the musical sound. In the double-reed instruments the vibrating reeds are coupled to a resonant air column which has the shape of a conical tube. A conical air column behaves like an open cylindrical tube and also gives rise to a fundamental and higher vibrational modes that are the even and odd harmonics of the fundamental.

The modern pipe organ consists of many resonant tubes or pipes some of which are the flue pipes that operate on the edge-tone phenomenon to produce sound and some are of the reed-type. These are controlled by one or more manual keyboards and by a pedal keyboard. Most of the flue pipes are made of zinc, tin or some alloy, and the tuning of the open pipe is accomplished by rolling, up or down, a narrow strip cut in the pipe near the top. This effectively lengthens or shortens the pipe. For the flue pipes that are closed at the top tuning is accomplished by the insertion of a snug-fitting plug at the upper end. In the reed type of pipe organ the reed is of thin metal and its natural frequency is made nearly equal to that of the pipe. Two types of reeds have been employed in organs; the free reed which is narrow and does not come in contact with the slot in the pipe, and the striking or beating reed in which the reed is somewhat larger than the slot or aperture over which it vibrates and therefore beats against the opening. Since the beating reed is easier to tune and maintain if tuned it is used almost exclusively today in organs. In general the pipes of the organ may be grouped according to their tone quality as diapasons, flutes, strings, and reeds. The closed pipes emit mostly the odd harmonics, and the open pipes possess both the even and the odd harmonics. The diapason, which is a cylindrical metal open pipe, is the most common pipe and its tone gives the organ its basic tonal quality.

Lip-reed and Vocal-cord Instruments

In the lip-reed brass wind instruments the lips of a player interrupt the steady air stream provided by the lungs and function as a mechanical double membranous reed which serves as the sound generating vibratile element. The stream of air against tightly drawn lips across a conical or cup-shaped mouth piece causes the lips to vibrate at a frequency that is a function of their tension, their length, and the amount of pressure. The steady air stream is throttled, by the vibration of the lips, into a pulsating one whose wave shape is approximately of the saw-tooth type that contains the even and the odd harmonics. This is introduced into the resonant column of the instrument with the consequent production of standing waves and sound. The common lip-reed instruments are the bugle, the trumpet, the cornet, the French horn, the trombone and the tuba. each of these has its own characteristic quality and differs in significant constructional features. For example, the cupped mouthpiece of the bugle is coupled to a coiled tube that ends in a flare or bell-shaped mouth and has a resonant length that is fixed. Therefore the number of notes that can be sounded is very limited. In the trumpet, cornet, French horn and tuba there are valves that permit the insertion of additional short pieces of tubing to increase the overall length of the resonant air column. In the slide trombone there is a U-shaped telescoping cylindrical tube that permits the lengthening of the resonant air column over relatively wider limits. The mouthpiece, the resonant tube and the bell cooperate to make the lip-reed brass instruments function as an open pipe. The shape of the mouthpiece produces a decided effect on the tone quality; a conical mouthpiece functioning to produce a smooth tone and a cup-shaped mouthpiece aiding in the production of higher upper partials and a more brilliant tone.

In the human voice the generators of sound are the two *vocal cords* which are membranous tissues that act like musical strings. They perform the function of a double reed and are set into vibration by streams of air that come from the lungs. The vocal cords are controlled by muscles that can change their length, their thickness, and their tension each of which plays a part in altering the frequency of vibration and of the emitted sound. The vocal resonance cavities are in the mouth, the nose, the lungs and the throat whose sizes are regulated by the tongue, the lower jaw, the lips and the cheeks, the chest, and the diaphragm. The general wave shape of the output of the vocal cords is approximately saw-tooth and contains a fundamental and all the harmonics. The adjustments that the human voice can effect in pitch, loudness and quality is unequaled by any other instrument.

8.3. The Family of Percussion Instruments

In Section 5.4 we discussed the vibrations of rods, bars, plates and membranes all of which are in the class of percussion instruments in which the vibrations are excited by striking the vibrating element. As musical instruments there are two percussion groups. In the one there are those instruments that possess a definite frequency or pitch such as the tuning fork, the xylophone and marimba, the glockenspiel, the celesta, the bells and carillon, the chimes and the kettledrums (tympani). In the other there are the instruments that do not possess a definite frequency or pitch such as the drums (snare, bass, military), the gong, the triangle, the cymbals and the tambourine. Rods may vibrate in longitudinal, torsional, and transverse modes. The vibrations of plates and membranes are two-dimensional vibrators. A bell is related to a plate and may be looked upon as a plate suspended from its center and curved downward. The fundamental frequency of a bell is a function of its internal diameter, its wall thickness, and some physical properties of the material. The pitch of a kettledrum is varied by changing the tension of the membrane by means of screws situated on its circular extremity. The bass drum does not have a definite pitch and is employed to provide rhythm and special effects. The tones of percussion instruments are highly transient and die out quickly after being produced. Most tones of percussion instruments are not periodic and their partials are not harmonic.

8.4. Electronic Instruments

Mechanical vibrations can be faithfully converted into electrical oscillatory currents or voltages if the metal vibratile element, such as a string, is surrounded by a magnetic field. Conversely, a wire carrying an electrical current and situated in a magnetic field experiences a mechanical force that tends to move it if the current is changing or alternating or if the magnetic field is changing or alternating. These two effects are the physical basis of the operation of *transducers* that are capable of transforming mechanical vibratory motions into alternating electric currents or voltages and alternating electric currents or voltages into mechanical vibrations. The microphone is an example of a sound transducer that responds to sound waves and converts them to electrical oscillations. The loudspeaker is an example of a transducer that converts the electrical oscillations into mechanical vibrations that produce sound. We shall not go into the various kinds of transducers that operate on different electrical principles since that would take us too far afield in the development of electrical principles. For our purposes it is sufficient to indicate that modern electronic developments have made it feasible to produce highly sophisticated transducers of such small size as to permit their use with the vibrations of strings, reeds, and other sound-producing elements. Such transducers are employed in various electronic musical instruments such as the electrical guitar, the electronic organ, the electronic piano, and the electronic carillon.

In the electrical guitar the vibrations of any of the six strings may be communicated to the bridge whose motion is transmitted to a small mechanoelectric transducer that converts the mechanical vibrations into corresponding electrical oscillations, or the vibrations may be likewise sensed directly by the transducer. The electrical variations are in turn amplified by an electronic amplifier whose output is coupled to a loudspeaker that converts the electrical variations into corresponding sound vibrations. Equipped with a volume control the electrical guitar may be adjusted to a wide range of sound levels. Similar transducers can be attached to violins, violas and other stringed instruments.

In one form of the electrical organ each note is produced by an alternator arrangement that consists of a tone wheel that is mounted to rotate in front of a magnet around which a pick-up coil of insulated wire is wound as shown in Figure 8.2. The periphery of the tone wheel disk of soft iron is shaped with a specific wave form. As it rotates it causes the magnetic flux to change in accordance with the undulations in its edge and this induces, by electromagnetic action, corresponding current or voltage changes in the pick-up coil. The outputs of many such generators are mixed in any desired proportion, amplified electronically, and converted into sound waves by a loudspeaker. The speed of rotation of each tone wheel, along with the wave shape at its periphery, determines the frequency that is generated. It is thus possible to synthesize any combination of harmonics and partials to simulate the various tones of a pipe organ. This tone-wheel method of synthesis is incorporated in the well-known electronic organ.

In another form of electronic organ there are employed electronic oscillators as the generators of the tones. By this means any wave shape can be generated and mixed to produce any complex tone. With generators of this type it is possible to produce sinusoidal waveforms that contain only one frequency, saw-tooth waveforms that contain both the even and odd harmonics and square-shaped wave forms that contain only the odd harmonics. By proper mixing and electronic filtering any desired wave form may be obtained and the waveform of any musical instrument may be simulated.

There are also electronic organs with air-driven metallic reeds which form one plate of an electrical capacitor whose capacitance (electrical charge-storing ability) changes as the distance between a fixed plate and the vibrating reed changes thus converting the mechanical vibration into an electrical variation. This type of electrostatic generator, employing an electrostatic transducer, is capable of developing an electrical output that is rich in harmonics. By the use of electrical filtering any desired waveform may be obtained. Any sound that is generated directly by the vibration of the reed is suppressed to avoid any interference with the electronic generation.

In the electronic piano the sound board is replaced by an electrical system that contains electromagnetic transducers or electrostatic transducers, of the type described above, to convert the vibrations of the strings into electrical oscillations. The output of the transducer is amplified and then converted into the corresponding sound waves by a loudspeaker. The electronic piano can be made relatively small in size and can be made suitable for a large auditorium or for an average-sized living room.

The electrical carillon similarly employs electrostatic or electro-magnetic transducers to convert the vibrations of rods, bars, chimes, bells, etc. to corresponding electrical variations. These in turn are amplified and fed into a loudspeaker for conversion into sound. The hammers that are used to strike the vibrators are actuated by the keys that form a keyboard.

8.5 The Musical Sound Synthesizer

In the preceding section we have seen how a performer may synthesize musical sounds in an electronic organ by combining many tone generators to produce any desired wave form. Highly sophisticated and relatively complex machines have been developed for synthesizing any combination of tones to produce varieties of complex sounds. In one such electronic musical synthesizer* music is produced by a machine from a coded record that is first prepared by a composer or musician. The synthesizer possesses electronic components that provide a means for producing a tone with any frequency, intensity, rate of growth, duration and decay, quality or timbre, vibrato, and with continuous glide aspects as in the violin, trombone, etc., in short, a tone with all of the significant characteristics of a sound. The composer, in his quest to create music, prepares a paper record of his musical composition. The record consists of rows of holes or perforations that are punched by means of a keyboard punching system and arranged in binary code. The record bears all of the characteristics of a musical tone that the composer wishes to specify. After the various elements of the synthesizer have been set in accordance with the specified characteristics the record is fed through the machine and the synthesized result can either be heard immediately or may be recorded on disk or tape. There is also available as part of the overall synthesizer a system for combining the different recordings produced by the synthesizer. This kind of synthesizer can produce simulations of the voice and of all the musical instruments as well as new tones that the voice or existing instruments are unable to

*H.F. Olson and H. Belar, "Electronic Music Synthesis," Jour. Acoust. Soc. Amer., 27, 595, 1955.
*H.F. Olson and H. Belar and J. Timmens, "Electronic Music Synthesis," Jour. Acoust. Soc. Amer., 32, 311, 1960.

produce. Furthermore the composer need not confine himself to common and familiar sounds. He may employ various electronically produced waves such as sinusoidal waves, square waves and saw-tooth waves in conjunction with various electronic filters. Among the advantages of this kind of electronic musical synthesizer may be mentioned that a composer is enabled to produce and speedily experiment with music creation without needing the services of a performer, that new complex sounds and combinations of sounds may be produced and evaluated as to their musical satisfaction and gratification, and that very old recordings in poor condition may be analyzed and then synthesized to produce a recording without distortion and noise.

In the Moog synthesizer use is made of voltage controlled function generators, amplifiers and special filters with the result that a musician can control the synthesizer from a keyboard rather than the use of punched tape to control the synthesizer.

An electronic musical synthesizer can modify natural sounds and generate sounds that are ordinarily not present in the emission of conventional musical instruments. It can reproduce all acoustical effects and control them electronically by combining sinusoidal waves, square waves, saw-toothed waves and pulsed waves, all of which are generated electronically with a controllable range of frequencies from a few hundredths hertz to about 10^4 hertz. The electronic musical synthesizer therefore hardly has any physical limitations and is a new kind of medium of expression. Its potentialities have not been fully realized and its additional benefits to the musician, the composer, and the listener will become available as the ongoing research in this phase of musical acoustics continues.

Problems

8. 1.   Is there a difference in the quality of the sound emitted by a string when it is bowed and when it is plucked? Explain.

8. 2.   The strings of the lower register of a piano are made of more massive material than the material of the strings in the higher register. Why?

8. 3.   The open A string on a violin has a length of 32.0 cm. Where should the fingers be placed, as measured from the bridge, so that the bowed string will emit the major diatonic scale? Retain one decimal in your result.

8. 4. What is the function of a sounding board in a musical instrument?

8. 5. Other than those mentioned in the chapter cite examples of sounds generated by the edge tone phenomenon that you may have experienced hearing.

8. 6. Make a chart of the various families and members of the musical instruments.

CHAPTER 9

Acoustics of Halls and Auditoriums

In our treatment of the properties of waves (Section 3.4) we have indicated that sound waves reflect from a medium upon which they fall and the amount or degree of reflection depends upon the absorption properties of the medium. We have also seen that such resulting echoes become multiplied when the sound energy reflects successively from the different boundaries and obstructions in an enclosure. We also pointed out that certain geometrically-shaped surfaces, like highly reflective concave surfaces (ceiling domes, for example) are to be avoided since they have undesirable focusing properties, whereas such a surface as a stage back drop may project the sound to the audience. On the other hand, convex surfaces like cylindrical pillars, produce the desirable effect of diffusing the sound and aid in assuring a more constant intensity throughout an auditorium or hall. Such characteristics of sound reflections were ascertained by utilizing the concept of a ray of sound, analogous to a ray of light. However we have seen that the phenomenon of diffraction was most significant to a listener, relative to the effect of the bending of sound waves around obstacles and through apertures and that these effects were best considered by using the wave property of sound. This lingering of echoes in a room in which there is an audience gives rise to a reverberation whose consideration is of paramount importance in the analysis of what constitutes good acoustics for a concert hall or auditorium. We shall consider this more closely in the present chapter.

9.1. Reverberant Sound
The sustained echoing in an enclosure due to multiple reflections of

sound from marble, stone, concrete, wood, etc. results in a prolongation of the original sound and a diffuse mixture of sound called *reverberant sound*. This resultant scrambling and unintelligibility of speech and music is illustrated by the curves in Figure 9.1, in which the ordinate of the plot is sound intensity and the abscissa is the time. First consider the graph ABCD which illustrates the growth and decay characteristics of a musical sound emitted by a source in an enclosure. A sound that is sustained over a period of several seconds will grow in intensity as indicated by the exponential section AB and attain a plateau or steady state condition at B when the sound energy is being absorbed or generally dissipated at a rate equal to its rate of emission. If, at a time corresponding to point C, the source is made to cease its emission or sound, the sound intensity will gradually diminish or decay somewhat as indicated by the section CD also in an exponential manner. How quickly or how slowly it takes the sound to reach its equilibrium state on growth and how quickly or how slowly it takes the sound to "die out" so that it is inaudible depends on such factors as the size and shape of the enclosure, and the absorption characteristics of the interior surfaces. Since the decay curve is exponential it is true that after an interval of time known as the *half-life* the intensity has dropped to half its value. Mathematically this means that the sound intensity attains a value of zero only after an infinite time lapse. However, for all practical purposes, the sound intensity is inaudible, and we consider it to be zero, after it has decreased by 60 dB from its original intensity. The *reverberation time*, or time interval of decay, is therefore defined as the time required for the sound at a given frequency to decay from a level 60 dB above the threshold of audibility or a decrease of one-millionth of the original sound intensity. The reverberation time of course is of principal significance in the evaluation of the acoustic qualities of a given hall or auditorium. Although we have drawn the curves in Figure 9.1 as smooth they represent the average of the actual vibrational wiggles or undulations.

If the reverberation time is too long, the emission of a speech syllable or a musical sound is mixed with the previous and still existing speech syllables or musical sounds and there results a disturbing unintelligibility of the speech or a condition of musical distortion. An enclosure with too long a reverberation time is termed too live. This condition for musical sounds is indicated in Figure 9.1 by the two additional growth and decay curves which are drawn to follow the initial curve ABCD before its sound intensity has decreased sufficiently. The cross-hatched areas represent the disturbing overlapping regions. For speech the growth part of the curves do not reach their steady state regions BC and the decay portion CD commences at some time between A and B. If the reverberation time is too short, there is too much absorption, resulting in a hall or auditorium that is too dead and there is not enough reinforcement of the original sound. In fact, for accurate acoustic measurements there are specially constructed rooms called *anechoic chambers* that are built to absorb all of the sound incident on its enclosing boundaries.

Obviously there must be an optimum reverberation time for a given room that is to be used for a given purpose. Such optimum reverberation times may be determined experimentally for various sized and shaped enclosures that may be used for various purposes by utilizing the emission of single-frequency sound sources for music or by utilizing the emission of speech sounds. The measured quantity is the time it takes the sound to decay from a level of 60 dB above threshold of audibility to the threshold. Different techniques may be employed to obtain such results, some of which employ the subjective judgment of an observer while other methods eliminate to a large degree the errors of personal judgment. Since measurements of this nature depend upon the sound frequency, the size of the enclosure, the absorption characteristics within the enclosure including an audience, and whether the enclosure is to be used for speech, for music or for both, the published data on reverberation times by different researchers are not in agreement and must be used with extreme care. Representative frequencies that are ordinarily employed are 500, 1000, and 2000 Hz. Table 9.1 lists our compilation of optimum reverberation times at 500 Hz to be found in various research publications, where we have indicated a range in time values as a function of the volume of the enclosure and of the kind of sound for which the enclosure is primarily to be used. It is to be noticed that the entire spread for optimum time values is from about 0.5 second to about 2.2 seconds. The range of values indicated is approximate but may be employed for rough computation in the evaluation of the acoustic quality of a hall or auditorium or in the design of such enclosures, as described later in the chapter.

Table 9.1

Optimum Reverberation Time at 500 Hz as a Function of Enclosure Volume, and Type of Sound

Kind of Sound	Optimum Reverberation times in seconds at 500 Hz Volume V in thousands of cubic feet			
	V: less than 10	V: 10-100	V: 100-500	V:500-1000
Speech	0.5 - 0.7	0.7 - 1.0	1.0 - 1.3	1.3 - 1.4
Speech and Music	0.8 - 1.1	1.1 - 1.4	1.4 - 1.6	1.6 - 1.7
Theatre	0.8 - 1.1	1.1 - 1.4	1.4 - 1.8	1.8 - 2.0
Orchestra	—	1.3 - 1.6	1.6 - 1.9	1.9 - 2.0
Organ	—	1.2 - 1.6	1.6 - 2.0	2.0 - 2.2

9.2 Sabine's Reverberation Formula

In 1895 Professor W.C. Sabine, at Harvard University, began a study of the properties of an enclosure that determine how effective the enclosure is for speech and music. His thorough researches and findings established the science of architectural acoustics. The definition of the reverberation time stated in the preceding section is due to Sabine. After careful experimental investigations of the acoustical properties of an enclosure Sabine arrived at an empirical relationship between the natural reverberation time, the volume of the enclosure, and the amount of absorbing material in the enclosure. Sabine's formula or equation, theoretically deduced and experimentally verified by him, for the referberation time T is

$$T = \frac{0.049V}{A} \tag{9.1}$$

where T is in seconds, V is the volume of the enclosure in cubic feet, and A represents the total sound absorption by all the surfaces in the enclosure. In any hall or auditorium there are a number of different surfaces on which the sound falls and each surface absorbs sound by a different amount. Hence the factor A is a sum of terms given by

$$A = a_1 S_1 + a_2 S_2 + a_3 S_3 + --- \tag{9.2}$$

where S_1 is the total area in square feet of one kind of material in the enclosure and a_1 is the absorption coefficient of this material, S_2 is the total area in square feet of a different kind of material in the enclosure and a_2 is the absorption coefficient of that material, etc.* The *absorption coefficients* represent the fraction of the incident sound energy absorbed by the different materials and the products aS, expressed in square feet, represent the absorption of these materials. The unit of sound absorption aS has been named the *sabine* after Professor Sabine. Table 9.2 lists the absorption coefficients of some materials, and since these are a function of the frequency, being generally smaller the lower the frequency, they are indicated for a frequency of 500 Hz. The values given are only representative since the effective absorption depends also upon the thickness of the material, upon the manner in which the material is mounted, and upon the technique of measurement. The values are based on the stipulation that an absorption of one sabine represents a surface capable of absorbing sound at the same rate as one square foot of a perfectly absorbing surface. Such a surface would absorb 100 percent of the energy incident upon it and so have an absorption coefficient of unity, such as an open window. A material that has an absorption coefficient of say 0.4 will absorb 40 percent of sound energy falling upon it and reflect 60 percent and 25 square feet of such material will absorb as much as 10 square feet of a perfect absorber or have an absorption of 10

Table 9.2.
Sound Absorption Coefficients at 500 Hz of Various Materials

Material	Absorption Coefficient
Acoustic paneling	0.50
Acoustic plaster	0.50
Brick wall (unpainted)	0.03
Carpet (heavy)	0.37

*Sabine's formula is approximate and does not yield reliable results for enclosures that are very large (volumes greater than about 10^6 ft^3) nor for smaller rooms with highly absorbant walls. However for concert halls, auditoriums, theatres, etc. of usual dimensions it yields reliable data. The Eyring formula $T = 0.049$ V/-Sln $(1-\alpha)$ where $S = S_1 + S_2 + S_3$ — — —, ln is the natural logarithm, and $\alpha = A/S$, yields reliable results for $\alpha > 0.15$. The Sabine formula is good for $\alpha < 0.15$.

Concrete (unpainted)	0.01
Draperies (heavy)	0.50
Draperies (light)	0.10
Felt	0.56
Glass	0.04
Fiberglas (acoustical)	0.70
Fiber tile (perforated cellulose)	0.71
Marble	0.01
Plaster	0.05
Plywood	0.10
Rockwool	0.63
Tile (glazed)	0.01
Wood (hard, varnished)	0.03
Wood paneling (pine)	0.10
Audience in upholstered seats, per person (in sabines)	5.0
Upholstered seat (unoccupied)(in sabines)	3.5

sabines. The table illustrates that some materials such as marble, glazed tile and unpainted concrete absorb very little sound energy while other substances like acoustical fiberglass absorb as much as 70 percent of the sound energy falling on it. In general, most acoustical porous materials are good absorbers. When sound waves impinge upon such surfaces a good portion of the energy is trapped in the interstices of the material and is thus converted into heat energy. The result is as if this energy has escaped through an open window.

Equation 9.1, may be derived theoretically by employing the concept of geometric or ray acoustics. Here sound from a source, and after reflection from surfaces, in an enclosure is considered to travel outward in straight lines thus eventually producing the same average sound energy density throughout the enclosure. This procedure does not consider the presence of diffraction effects of sound waves particularly at the low frequencies, neglects the consideration of the normal modes of vibration or resonant frequencies whose persistence results in the reverberant sound, and does not take into account focusing effects due to the geometry of the enclosure and interference effects. Nevertheless, although Equation 9.1 is approximate, predicting values of T that are incorrect for large enclosures or for enclosures with large absorption, and therefore must be modified for more accurate diversified representation, it is sufficiently accurate for our purposes. The Sabine equation permits the determination of the reverberation time in an enclosure. It also indicates that the reverberation time in an enclosure can be reduced or increased by the insertion or removal respectively of absorption materials at the walls and other locations. The equation may also be employed to measure the change in reverberation time when a known area of absorbing material is introduced in a specially constructed enclosure known as a reverberation chamber and thereby determine the absorption coefficient of the material. Values of absorption coefficients so determined are however usually too large.

9.3. Computation of Reverberation Time

By utilizing the Sabine equation in conjunction with a knowledge of optimum reverberation times, such as are indicated in Table 9.1, it is possible to arrive at remedial measures that will bring the reverberation time of a hall or auditorium to a value that will make it more suitable for speech and/or musical rendition. We shall now illustrate such a procedure by considering as an example an auditorium whose length, width and height are respectively 60 feet, 30 feet and 20 feet. The floor, covered with varnished hardwood, is equipped with 120 upholstered seats spread over 85 percent of the floor area. The walls are brick, the ceiling is plaster, and the enclosure contains 30 square feet of glass window on each of the side walls and 15 square feet of glass window on one end wall. To compute the reverberation time for 500 Hz using the Sabine equation we tabularize the data needed to obtain the total absorption.

	Area (sq. ft.)	Abs. Coeff.	Abs. (sabines)
Side walls, less windows	2 x 20 x 60 - 60 = 2340	0.03	70.2
End walls, less window	2 x 20 x 30 - 15 = 1185	0.03	35.6
Ceiling	30 x 60 = 1800	0.05	90.0
Windows	75	0.04	3.0
Floor, less seat coverage	0.15 x 30 x 60 = 270	0.03	8.1
TOTAL			206.9
Seats, unoccupied	120	3.5 (sabines)	420.0
Seats, 1/2 occupied	60	5.0 (sabines)	300.0
Seats, all occupied	120	5.0 (sabines)	600.0

Volume of Enclosure = 20 x 30 x 60 = 36,000 cubic feet

Substituting in the Sabine formula,

$$T(\text{seats unoccupied}) = 0.049 \times 36,000/(207 + 420) = 2.8 \text{ sec.}$$
$$T(\text{seats } 1/2 \text{ occupied}) = 0.049 \times 36,000/(207 + 210 + 300) = 2.5 \text{ sec.}$$
$$T(\text{seats all occupied}) = 0.049 \times 36,000/(207 + 600) = 2.2 \text{ sec.}$$

Let us now assume that the auditorium was constructed to be used for speech and for music. Examining Table 9.1 we observe that the optimum reverberation time should be between 1.1 and 1.4 seconds, let us say 1.2 seconds. Since the computation above indicates that even with a full audience the reverberation time is too high we are guided by the application of the Sabine formula to insert additional sound-absorbing material. We may determine how much surface of sound-absorbing material to introduce by setting $T = 1.2$ seconds in the Sabine formula and solving for the total absorption required. This yields the value

$$A = 0.049 \times 36000/1.2 = 1470 \text{ sabines}$$

Considering remedying the acoustics of the auditorium when half occupied by an audience, the additional units of absorption needed to give a reverberation time of 1.2 seconds is $1470 - 717 = 753$ sabines. If we consider introducing fiber tile with an absorption coefficient of 0.71 the required amount of surface area of this material may be roughly obtained by dividing 753 by 0.71 which gives 1061 square feet. However, if we consider covering the walls or the ceiling, this is too low a value since the tile placed on the walls or ceiling will cover the brick or plaster and this amount of brick or plaster will no more absorb sound. In the interest of simplifying the solution of the problem we shall consider covering the ceiling although for small halls and auditoriums reflections from the ceiling are needed to carry the sound to the rear of the enclosure. We can obtain the surface area to be covered by letting x be the fraction of the ceiling area that remains plaster and (1-x) be the fraction that is covered with fiber tile and again use the Sabine equation to yield 1.2 seconds reverberation time. We thus have

$$1.2 = \frac{(0.049)\ (36000)}{627 + (1800)(0.05x) + (1800)(0.71)(1\text{-}x)}$$

and solving for x we obtain

$$x = 0.37$$
$$(1\text{-}x) = 0.63$$

Therefore 63 percent of 1134 square feet of the ceiling needs to be covered with the absorbing fiber to obtain the suitable reverberation time. A similar calculation can be made if it is desirable to cover the walls instead of the ceiling (see Problem 9.2).

9.4. The Design of Halls or Auditoriums

In the preceding section we have shown how it is possible and relatively simple to determine corrective measures to bring the reverberation time of a hall or auditorium to a value that will make the enclosure exhibit more suitable acoustical properties. For the reasons we have already indicated this procedure using the Sabine equation yields results that are approximate and at best a guide to remedial methods that may be applied to rooms, halls, and auditoriums that do not possess good acoustical qualities. The process of designing a hall or auditorium, prior to construction, so as to insure that it possesses good acoustical properties for speech or musical-rendition purposes, is a much more difficult task. The acoustical properties of an enclosure are good if the sound has sufficient intensity in every part of the enclosure, particularly under balconies in an auditorium, if all of the components of a complex sound retain their original relative intensities, if the reverberation time is neither too long nor too short with the respective attendant overlapping or dead disturbing qualities, and if the sound to be heard is free from internal or external extraneous noises. The correct acoustical design of an enclosure intended for audience purposes involves not only the volume of the enclosure, the sizes and shapes of the contours of the walls and ceilings, and the physical nature of these surfaces but also the effects of the interference and diffraction of the sound waves, the dependence of the reverberation time on the frequency of the sound (absorption coefficients are highly frequency dependent), the dependence on the absorption of the air and its temperature and humidity, the avoidance of focused reflections that cause intense discrete echoes, an acoustic provision that enables members of an orchestra or ensemble to hear one another clearly and distinctly, and the manner in which a human listener actually perceives the sound and just what characteristics of the sound the listener wishes to hear. Remedies for some of the aspects that play a significant role in providing satisfactory acoustics are continually being provided. For example, in the design of modern concert halls and auditoriums, the seats are constructed and upholstered in a manner that results in each having an absorption coefficient that is the same as that of an adult person occupying the seat. As another example, halls in which two opposite sides are flat and parallel are avoided in design in order to eliminate multiple reflections back and forth and so avoid the undesirable rapidly repeating echo known as *flutter echo*. Also, external noises from traffic, airplanes, etc. can largely be excluded by proper sound insulation during construction. However much still remains to be done not only with remedial techniques but also, and especially so, in the realm of prediction before a hall or auditorium is constructed. Methods that are yielding promising results in this respect are currently being employed. In one,* a model of the proposed hall or auditorium is built to scale and sounds of the appropriate short wavelength are produced inside the model. The results of the studies of such models, including the interference and diffraction effects, lead to a knowledge of the deficiencies and to appropriate remedial measures before construction. In another method,** which appears to possess a high degree of predictability, an electronic computer is employed to simulate the acoustics of existing and proposed halls or auditoriums using the method of modeling. The appropriate computer program, based on information from architectural plans, is used to modify a tape recording of music in such a way as to sound as if it were occurring in the hall or auditorium. The sound is studied and evaluated by stereophonically playing the tape back through a group of loudspeakers in an anechoic chamber.

The subject of hall and auditorium acoustics is being understood more and more as researches in this field continually uncover significant developments. Perhaps in time the reliability of predicting the design of an enclosure with excellent acoustic qualities will improve to such an extent as to reduce to a considerable degree the practice of first building the hall or auditorium and then calling in the acoustical expert to remedy the deficiencies.

*E. Krauth and R. Buchlein, Model Tests in Architectural Acoustics, Gravesano Review, 27/28, 155-60, 1966.
**M.R. Schroeder, Computer Models for Concert Hall Acoustics, Am. Jnl. Phys., *41*, 461, 1973.

Problems

9. 1.   Roughly draw a series of curves like those shown in Figure 9.1 that illustrate the overlapping regions for speech.

9. 2.   In the example given in Section 9.3 compute the area of brick wall to be covered by the fibre tile rather than the ceiling.

9. 3.   A large rectangularly-shaped room has the dimensions of width 120 feet, length 160 feet, and height 40 feet. The absorption coefficients are 0.1 for the ceiling, 0.2 for the side walls, 0.3 for the front and rear walls and 0.05 for the floor. Find (a) the total number of absorption units and (b) the reverberation time of the room.

9. 4.   (a) How much area of the ceiling in Problem 9.3 needs to be covered with acoustic material of absorption coefficient 0.9 to bring the reverberation time down to 2 seconds? (b) Can this reverberation time be obtained by covering the side walls of the room of the ceiling?

9. 5.   An auditorium whose length, width and height are respectively 80 feet, 60 feet and 30 feet is being used for speech and music purposes. The floor, covered with varnished wood, is equipped with 150 upholstered seats spread over 80 percent of the floor area. The walls and ceiling are plaster. Find how much area of the walls, and of the ceiling if necessary, must be covered with acoustic tile having an absorption coefficient of 0.8 in order that the enclosure have an appropriate reverberation time (use the average of the range given in Table 9.1). Make the computation corresponding to the seats being two-thirds occupied.

9. 6.   A band shell is usually present in back of the musicians that are participating in a band concert in a park where the seats or benches are on ground covered with grass. Can you give a reason for the presence of the shell?

9. 7.   For large auditoriums, as well as for out-door performances, electronic amplification is usually necessary. If loudspeakers are mounted out over the audience, what precautions must be employed to insure clear sound reception without successive repetition of speech syllables or musical sound phrases?

APPENDIX
Table A.1. *Natural sines and cosines.*

Sines (read down)   **Cosines (read up)**

	.0	.1	.2	.3	.4	.5	.6	.7	.8	.9	
0°	.0000	.0017	.0035	.0052	.0070	.0087	.0105	.0122	.0140	.0157	.0175	89°
1°	.0175	.0192	.0209	.0227	.0244	.0262	.0279	.0297	.0314	.0332	.0349	88°
2°	.0349	.0366	.0384	.0401	.0419	.0436	.0454	.0471	.0488	.0506	.0523	87°
3°	.0523	.0541	.0558	.0576	.0593	.0610	.0628	.0645	.0663	.0680	.0698	86°
4°	.0698	.0715	.0732	.0750	.0767	.0785	.0802	.0819	.0837	.0854	.0872	85°
5°	.0872	.0889	.0906	.0924	.0941	.0958	.0976	.0993	.1011	.1028	.1045	84°
6°	.1045	.1063	.1080	.1097	.1115	.1132	.1149	.1167	.1184	.1201	.1219	83°
7°	.1219	.1236	.1253	.1271	.1288	.1305	.1323	.1340	.1357	.1374	.1392	82°
8°	.1392	.1409	.1426	.1444	.1461	.1478	.1495	.1513	.1530	.1547	.1564	81°
9°	.1564	.1582	.1599	.1616	.1623	.1650	.1668	.1685	.1702	.1719	.1736	80°
10°	.1736	.1754	.1771	.1788	.1805	.1822	.1840	.1857	.1874	.1891	.1908	79°
11°	.1908	.1925	.1942	.1959	.1977	.1994	.2011	.2028	.2045	.2062	.2079	78°
12°	.2079	.2096	.2113	.2130	.2147	.2164	.2181	.2198	.2215	.2233	.2250	77°
13°	.2250	.2267	.2284	.2300	.2317	.2334	.2351	.2368	.2385	.2402	.2419	76°
14°	.2419	.2436	.2453	.2470	.2487	.2504	.2521	.2538	.2554	.2571	.2588	75°
15°	.2588	.2605	.2622	.2639	.2656	.2672	.2689	.2706	.2723	.2740	.2756	74°
16°	.2756	.2773	.2790	.2807	.2823	.2840	.2857	.2874	.2890	.2907	.2924	73°
17°	.2924	.2940	.2957	.2974	.2990	.3007	.3024	.3040	.3057	.3074	.3090	72°
18°	.3090	.3107	.3123	.3140	.3156	.3173	.3190	.3206	.3223	.3239	.3256	71°
19°	.3256	.3272	.3289	.3305	.3322	.3338	.3355	.3371	.3387	.3404	.3420	70°
20°	.3420	.3437	.3453	.3469	.3486	.3502	.3518	.3535	.3551	.3567	.3584	69°
21°	.3584	.3600	.3616	.3633	.3649	.3665	.3681	.3697	.3714	.3730	.3746	68°
22°	.3746	.3762	.3778	.3795	.3811	.3827	.3843	.3859	.3875	.3891	.3907	67°
23°	.3907	.3923	.3939	.3955	.3971	.3987	.4003	.4019	.4035	.4051	.4067	66°
24°	.4067	.4083	.4099	.4115	.4131	.4147	.4163	.4179	.4195	.4210	.4226	65°
25°	.4226	.4242	.4258	.4274	.4289	.4305	.4321	.4337	.4352	.4368	.4384	64°
26°	.4384	.4399	.4415	.4431	.4446	.4462	.4478	.4493	.4509	.4524	.4540	63°
27°	.4540	.4555	.4571	.4586	.4602	.4617	.4633	.4648	.4664	.4679	.4695	62°
28°	.4695	.4710	.4726	.4741	.4756	.4772	.4787	.4802	.4818	.4833	.4848	61°
29°	.4848	.4863	.4879	.4894	.4909	.4924	.4939	.4955	.4970	.4985	.5000	60°
30°	.5000	.5015	.5030	.5045	.5060	.5075	.5090	.5105	.5120	.5135	.5150	59°
31°	.5150	.5165	.5180	.5195	.5210	.5225	.5240	.5255	.5270	.5284	.5299	58°
32°	.5299	.5314	.5329	.5344	.5358	.5373	.5388	.5402	.5417	.5432	.5446	57°
33°	.5446	.5461	.5476	.5490	.5505	.5519	.5534	.5548	.5563	.5577	.5592	56°
34°	.5592	.5606	.5621	.5635	.5650	.5664	.5678	.5693	.5707	.5721	.5736	55°
35°	.5736	.5750	.5764	.5779	.5793	.5807	.5821	.5835	.5850	.5864	.5878	54°
36°	.5878	.5892	.5906	.5920	.5934	.5948	.5962	.5976	.5990	.6004	.6018	53°
37°	.6018	.6032	.6046	.6060	.6074	.6088	.6101	.6115	.6129	.6143	.6157	52°
38°	.6157	.6170	.6184	.6198	.6211	.6225	.6239	.6252	.6266	.6280	.6293	51°
39°	.6293	.6307	.6320	.6334	.6347	.6361	.6374	.6388	.6401	.6414	.6428	50°
40°	.6428	.6441	.6455	.6468	.6481	.6494	.6508	.6521	.6534	.6547	.6561	49°
41°	.6561	.6574	.6587	.6600	.6613	.6626	.6639	.6652	.6665	.6678	.6691	48°
42°	.6691	.6704	.6717	.6730	.6743	.6756	.6769	.6782	.6794	.6807	.6820	47°
43°	.6820	.6833	.6845	.6858	.6871	.6884	.6896	.6909	.6921	.6934	.6947	46°
44°	.6974	.6959	.6972	.6984	.6997	.7009	.7022	.7034	.7046	.7059	.7071	45°

| | .9 | .8 | .7 | .6 | .5 | .4 | .3 | .2 | .1 | .0 | |

Table A.1. *(Continued)*

Sines (read down)   Cosines (read up)

	.0	.1	.2	.3	.4	.5	.6	.7	.8	.9	
45°	.7071	.7083	.7096	.7108	.7120	.7133	.7145	.7157	.7169	.7181	.7193	44°
46°	.7193	.7206	.7218	.7230	.7242	.7254	.7266	.7278	.7290	.7302	.7314	43°
47°	.7314	.7325	.7337	.7349	.7361	.7373	.7385	.7396	.7408	.7420	.7431	42°
48°	.7431	.7443	.7455	.7466	.7478	.7490	.7501	.7513	.7524	.7536	.7547	41°
49°	.7547	.7559	.7570	.7581	.7593	.7604	.7615	.7627	.7638	.7649	.7660	40°
50°	.7660	.7672	.7683	.7694	.7705	.7716	.7727	.7738	.7749	.7760	.7771	39°
51°	.7771	.7782	.7793	.7804	.7815	.7826	.7837	.7848	.7859	.7869	.7880	38°
52°	.7880	.7891	.7902	.7912	.7923	.7934	.7944	.7955	.7965	.7976	.7986	37°
53°	.7986	.7997	.8007	.8018	.8028	.8039	.8049	.8059	.8070	.8080	.8090	36°
54°	.8090	.8100	.8111	.8121	.8131	.8141	.8151	.8161	.8171	.8181	.8192	35°
55°	.8192	.8202	.8211	.8221	.8231	.8241	.8251	.8261	.8271	.8281	.8290	34°
56°	.8290	.8300	.8310	.8320	.8329	.8339	.8348	.8358	.8368	.8377	.8387	33°
57°	.8387	.8396	.8406	.8415	.8425	.8434	.8443	.8453	.8462	.8471	.8480	32°
58°	.8480	.8490	.8499	.8508	.8517	.8526	.8536	.8545	.8554	.8563	.8572	31°
59°	.8572	.8581	.8590	.8599	.8607	.8616	.8625	.8634	.8643	.8652	.8660	30°
60°	.8660	.8669	.8678	.8686	.8695	.8704	.8712	.8721	.8729	.8738	.8746	29°
61°	.8746	.8755	.8763	.8771	.8780	.8788	.8796	.8805	.8813	.8821	.8829	28°
62°	.8829	.8838	.8846	.8854	.8862	.8870	.8878	.8886	.8894	.8902	.8910	27°
63°	.8910	.8918	.8926	.8934	.8942	.8949	.8957	.8965	.8973	.8980	.8988	26°
64°	.8988	.8996	.9003	.9011	.9018	.9026	.9033	.9041	.9048	.9056	.9063	25°
65°	.9063	.9070	.9078	.9085	.9092	.9100	.9107	.9114	.9121	.9128	.9135	24°
66°	.9135	.9143	.9150	.9157	.9164	.9171	.9178	.9184	.9191	.9198	.9205	23°
67°	.9205	.9212	.9219	.9225	.9232	.9239	.9245	.9252	.9259	.9265	.9272	22°
68°	.9272	.9278	.9285	.9291	.9298	.9304	.9311	.9317	.9323	.9330	.9336	21°
69°	.9336	.9342	.9348	.9354	.9361	.9367	.9373	.9379	.9385	.9391	.9397	20°
70°	.9397	.9403	.9409	.9415	.9421	.9426	.9432	.9438	.9444	.9449	.9455	19°
71°	.9455	.9461	.9466	.9472	.9478	.9483	.9489	.9494	.9500	.9505	.9511	18°
72°	.9511	.9516	.9521	.9527	.9532	.9537	.9542	.9548	.9553	.9558	.9563	17°
73°	.9563	.9568	.9573	.9578	.9583	.9588	.9593	.9598	.9603	.9608	.9613	16°
74°	.9613	.9617	.9622	.9627	.9632	.9636	.9641	.9646	.9650	.9655	.9659	15°
75°	.9659	.9664	.9668	.9673	.9677	.9681	.9686	.9690	.9694	.9699	.9703	14°
76°	.9703	.9707	.9711	.9715	.9720	.9724	.9728	.9732	.9736	.9740	.9744	13°
77°	.9744	.9748	.9751	.9755	.9759	.9763	.9767	.9770	.9774	.9778	.9781	12°
78°	.9781	.9785	.9789	.9792	.9796	.9799	.9803	.9806	.9810	.9813	.9816	11°
79°	.9816	.9820	.9823	.9826	.9829	.9833	.9836	.9839	.9842	.9845	.9848	10°
80°	.9848	.9851	.9854	.9857	.9860	.9863	.9866	.9869	.9871	.9874	.9877	9°
81°	.9877	.9880	.9882	.9885	.9888	.9890	.9893	.9895	.9898	.9900	.9903	8°
82°	.9903	.9905	.9907	.9910	.9912	.9914	.9917	.9919	.9921	.9923	.9925	7°
83°	.9925	.9928	.9930	.9932	.9934	.9936	.9938	.9940	.9942	.9943	.9945	6°
84°	.9945	.9947	.9949	.9951	.9952	.9954	.9956	.9957	.9959	.9960	.9962	5°
85°	.9962	.9963	.9965	.9966	.9968	.9969	.9971	.9972	.9973	.9974	.9976	4°
86°	.9976	.9977	.9978	.9979	.9980	.9981	.9982	.9983	.9984	.9985	.9986	3°
87°	.9986	.9987	.9988	.9989	.9990	.9990	.9991	.9992	.9993	.9993	.9994	2°
88°	.9994	.9995	.9995	.9996	.9996	.9997	.9997	.9998	.9998	.9998	.9998	1°
89°	.9998	.9999	.9999	.9999	.9999	1.0000	.1.0000	1.0000	1.0000	1.0000	1.0000	0°
	.9	.8	.7	.6	.5	.4	.3	.2	.1	.0		

Table A.2. LOGARITHMS

N	0	1	2	3	4	5	6	7	8	9
10	0000	0043	0086	0128	0170	0212	0253	0294	0334	0374
11	0414	0453	0492	0531	0569	0607	0645	0682	0719	0755
12	0792	0828	0864	0899	0934	0969	1004	1038	1072	1106
13	1139	1173	1206	1239	1271	1303	1335	1367	1399	1430
14	1461	1492	1523	1553	1584	1614	1644	1673	1703	1732
15	1761	1790	1818	1847	1875	1903	1931	1959	1987	2014
16	2041	2068	2095	2122	2148	2175	2201	2227	2253	2279
17	2304	2330	2355	2380	2405	2430	2455	2480	2504	2529
18	2553	2577	2601	2625	2648	2672	2695	2718	2742	2765
19	2788	2810	2833	2856	2878	2900	2923	2945	2967	2989
20	3010	3032	3054	3075	3096	3118	3139	3160	3181	3201
21	3222	3243	3263	3284	3304	3324	3345	3365	3385	3404
22	3424	3444	3464	3483	3502	3522	3541	3560	3579	3598
23	3617	3636	3655	3674	3692	3711	3729	3747	3766	3784
24	3802	3820	3838	3856	3874	3892	3909	3927	3945	3962
25	3979	3997	4014	4031	4048	4065	4082	4099	4116	4133
26	4150	4166	4183	4200	4216	4232	4249	4265	4281	4298
27	4314	4330	4346	4362	4378	4393	4409	4425	4440	4456
28	4472	4487	4502	4518	4533	4548	4564	4579	4594	4609
29	4624	4639	4654	4669	4683	4698	4713	4728	4742	4757
30	4771	4786	4800	4814	4829	4843	4857	4871	4886	4900
31	4914	4928	4942	4955	4969	4983	4997	5011	5024	5038
32	5051	5065	5079	5092	5105	5119	5132	5145	5159	5172
33	5185	5198	5211	5224	5237	5250	5263	5276	5289	5302
34	5315	5328	5340	5353	5366	5378	5391	5403	5416	5428
35	5441	5453	5465	5478	5490	5502	5514	5527	5539	5551
36	5563	5575	5587	5599	5611	5623	5635	5647	5658	5670
37	5682	5694	5705	5717	5729	5740	5752	5763	5775	5786
38	5798	5809	5821	5832	5843	5855	5866	5877	5888	5899
39	5911	5922	5933	5944	5955	5966	5977	5988	5999	6010
40	6021	6031	6042	6053	6064	6075	6085	6096	6107	6117
41	6128	6138	6149	6160	6170	6180	6191	6201	6212	6222
42	6232	6243	6253	6263	6274	6284	6294	6304	6314	6325
43	6335	6345	6355	6365	6375	6385	6395	6405	6415	6425
44	6435	6444	6454	6464	6474	6484	6493	6503	6513	6522
45	6532	6542	6551	6561	6571	6580	6590	6599	6609	6618
46	6628	6637	6646	6656	6665	6675	6684	6693	6702	6712
47	6721	6730	6739	6749	6758	6767	6776	6785	6794	6803
48	6812	6821	6830	6839	6848	6857	6866	6875	6884	6893
49	6902	6911	6920	6928	6937	6946	6955	6964	6972	6981
50	6990	6998	7007	7016	7024	7033	7042	7050	7059	7067
51	7076	7084	7093	7101	7110	7118	7126	7135	7143	7152
52	7160	7168	7177	7185	7193	7202	7210	7218	7226	7235
53	7243	7251	7259	7267	7275	7284	7292	7300	7308	7316
54	7324	7332	7340	7348	7356	7364	7372	7380	7388	7396

Table A.2. LOGARITHMS (*Continued*)

N	0	1	2	3	4	5	6	7	8	9
55	7404	7412	7419	7427	7435	7443	7451	7459	7466	7474
56	7482	7490	7497	7505	7513	7520	7528	7536	7543	7551
57	7559	7566	7574	7582	7589	7597	7604	7612	7619	7627
58	7634	7642	7649	7657	7664	7672	7679	7686	7694	7701
59	7709	7716	7723	7731	7738	7745	7752	7760	7767	7774
60	7782	7789	7796	7803	7810	7818	7825	7832	7839	7846
61	7853	7860	7868	7875	7882	7889	7896	7903	7910	7917
62	7924	7931	7938	7945	7952	7959	7966	7973	7980	7987
63	7993	8000	8007	8014	8021	8028	8035	8041	8048	8055
64	8062	8069	8075	8082	8089	8096	8102	8109	8116	8122
65	8129	8136	8142	8149	8156	8162	8169	8176	8182	8189
66	8195	8202	8209	8215	8222	8228	8235	8241	8248	8254
67	8261	8267	8274	8280	8287	8293	8299	8306	8312	8319
68	8325	8331	8338	8344	8351	8357	8363	8370	8376	8382
69	8388	8395	8401	8407	8414	8420	8426	8432	8439	8445
70	8451	8457	8463	8470	8476	8482	8488	8494	8500	8506
71	8513	8519	8525	8531	8537	8543	8549	8555	8561	8567
72	8573	8579	8585	8591	8597	8603	8609	8615	8621	8627
73	8633	8639	8645	8651	8657	8663	8669	8675	8681	8686
74	8692	8698	8704	8710	8716	8722	8727	8733	8739	8745
75	8751	8756	8762	8768	8774	8779	8785	8791	8797	8802
76	8808	8814	8820	8825	8831	8837	8842	8848	8854	8859
77	8865	8871	8876	8882	8887	8893	8899	8904	8910	8915
78	8921	8927	8932	8938	8943	8949	8954	8960	8965	8971
79	8976	8982	8987	8993	8998	9004	9009	9015	9020	9025
80	9031	9036	9042	9047	9053	9058	9063	9069	9074	9079
81	9085	9090	9096	9101	9106	9112	9117	9122	9128	9133
82	9138	9143	9149	9154	9159	9165	9170	9175	9180	9186
83	9191	9196	9201	9206	9212	9217	9222	9227	9232	9238
84	9243	9248	9253	9258	9263	9269	9274	9279	9284	9289
85	9294	9299	9304	9309	9315	9320	9325	9330	9335	9340
86	9345	9350	9355	9360	9365	9370	9375	9380	9385	9390
87	9395	9400	9405	9410	9415	9420	9425	9430	9435	9440
88	9445	9450	9455	9460	9465	9469	9474	9479	9484	9489
89	9494	9499	9504	9509	9513	9518	9523	9528	9533	9538
90	9542	9547	9552	9557	9562	9566	9571	9576	9581	9586
91	9590	9595	9600	9605	9609	9614	9619	9624	9628	9633
92	9638	9643	9647	9652	9657	9661	9666	9671	9675	9680
93	9685	9689	9694	9699	9703	9708	9713	9717	9722	9727
94	9731	9736	9741	9745	9750	9754	9759	9763	9768	9773
95	9777	9782	9786	9791	9795	9800	9805	9809	9814	9818
96	9823	9827	9832	9836	9841	9845	9850	9854	9859	9863
97	9868	9872	9877	9881	9886	9890	9894	9899	9903	9908
98	9912	9917	9921	9926	9930	9934	9939	9943	9948	9952
99	9956	9961	9965	9969	9974	9978	9983	9987	9991	9996

Answers to Odd-Numbered Problems

Chapter 1

1.3. 5.00 cm,   53.1°.
1.7. (a) 88.0 ft/sec,   (b) 1.76 X 10^3cm.
1.11. (a) 1.75 km/sec,   (b) 8.75 km.
1.17. (a) 5.50 X 10^2 joules,   (b) 27.5 watts.

Chapter 2

2.3. 4.90 X 10^3 dynes/cm.
2.5. 5.00 X 10^{-2} sec/cycle to 5.56 X 10^{-6} sec/cycle.
2.7. $y = A \sin 2\pi ft + \frac{A}{2} \sin 2(2\pi ft)$.

Chapter 3

3.3. 4.98 X 10^3 m/sec.
3.5. 332 m/sec.
3.9. 69.0 cm.
3.11. ¾ cycle or 270°
3.13. 688 m.

Chapter 4

4.1. 2 units.
4.5. 450 m/sec.

Chapter 5

5.5. 2.74 N.
5.7. 4.17 m.
5.13. 90.0 cm.
5.15. 31.6 cm.

Chapter 6

6.5.   16.
6.11.   70 dB.
6.13.   60 dB.
6.15.   27 dB.
6.17.   90.4 dB.

Chapter 7

7.1.   5/4, major third; 5/3, major sixth; 3/2, fifth; 6/5, minor third.
7.3.   Dissonance due to beats between second upper partials and third upper partials.
7.5.   (a) 440 Hz, 275 Hz.   (b) 8/5, minor sixth.
7.7.   A, D, and E.
7.11.   24 cents.
7.13.   (a) 100.2 cents   (b) Pythagorean chromatic is nearer.

Chapter 8

8.3.   B, 28.8 cm;   C, 25.6 cm;   D, 24.5 cm;   E, 21.3 cm;
        F, 19.2 cm;   G, 17.1 cm;   A, 16.0 cm.

Chapter 9

9.3.   (a) 8320 sabines,   (b) 4.4 sec.
9.5.   4776 square feet of wall area.

Illustrations for
The Physical Basis
of
Musical Sounds

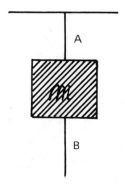

Figure 1.1. Problem 1.13.

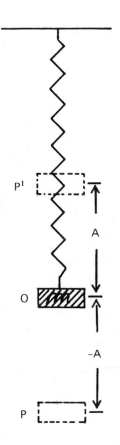

Figure 2.1. The simple harmonic motion of an oscillating mass suspended from an elastic spring.

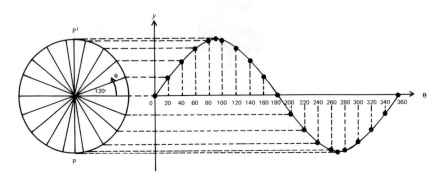

Figure 2.2. Graphical representation of SHM

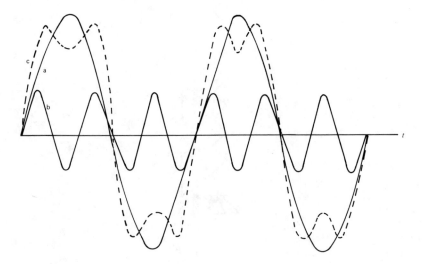

Figure 2.3. Simple waves combining to form a complex wave (dashed curve).

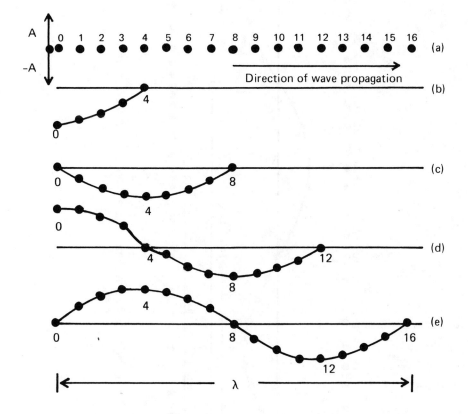

Figure 3.1. Mechanism of production of a transverse wave.

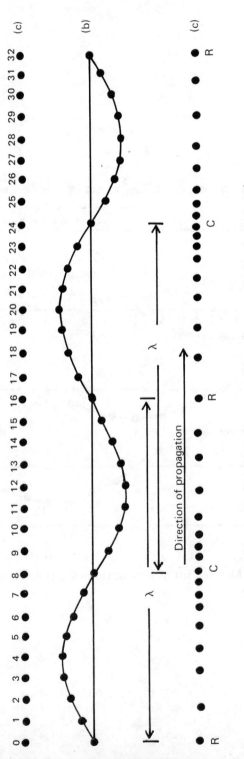

Figure 3.2. Longitudinal wave and its transverse wave representation

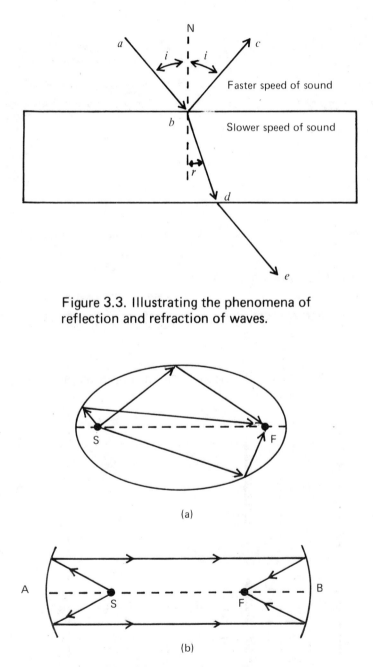

Figure 3.3. Illustrating the phenomena of reflection and refraction of waves.

(a)

(b)

Figure 3.4. Illustrating the concentration of sound energy to reflection from concave surfaces.

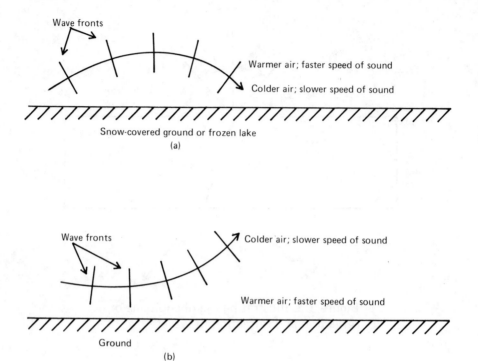

Wave fronts

Warmer air; faster speed of sound

Colder air; slower speed of sound

Snow-covered ground or frozen lake
(a)

Wave fronts

Colder air; slower speed of sound

Warmer air; faster speed of sound

Ground
(b)

Figure 3.5. Refraction of sound waves

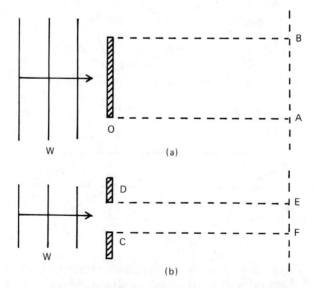

W

O

B

A

(a)

W

D

C

E

F

(b)

Figure 3.6. Diffraction by (a) an obstacle
and (b) an aperture.

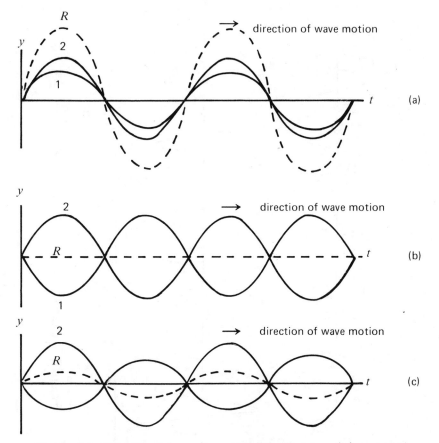

Figure 4.1. Showing (a) constructive interference, (b) complete destructive interference, and (c) partial destructive interference.

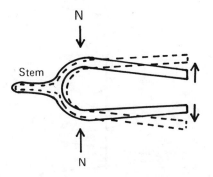

Figure 4.2. Vibration of a tuning fork.

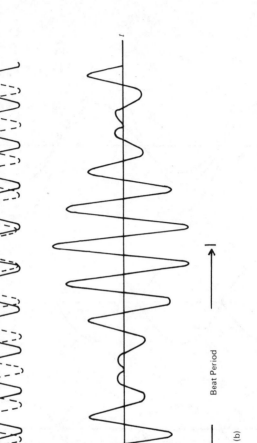

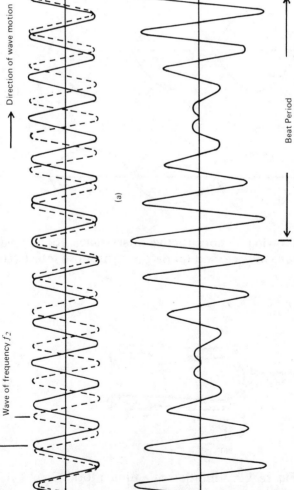

Figure 4.3. **Beats.**

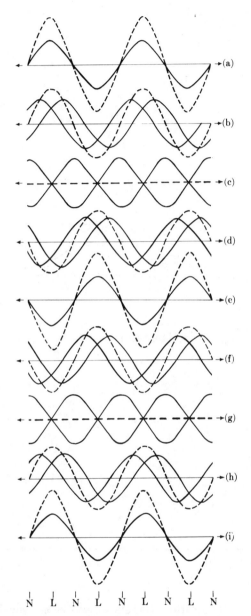

Figure 4.4. Illustrating the formation
of a standing wave. The oppositely directed
waves are the solid curves and the resultant
wave is shown by the dashed curve. The two
oppositvely directed waves are coincident in
phase in (a), (e) and (i). From (a) to (i) the
graphs successively differ by one eighth of
a period.

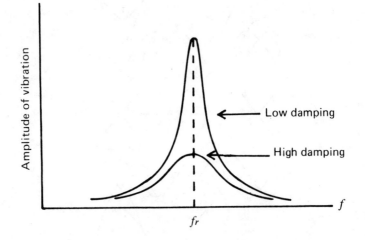

Figure 5.1. The resonance curve

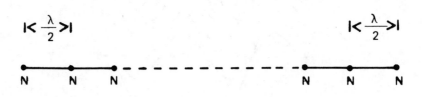

Figure 5.2. Illustrating the relation between wavelength and length of string for the formation of stationary waves and resonance.

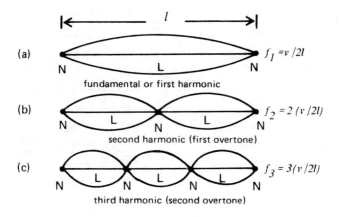

Figure 5.3. Vibrational modes of a string. Both even and odd harmonics are emitted.

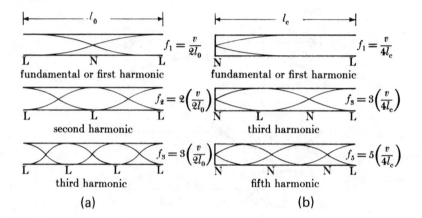

Figure 5.4. Vibrational modes of (a) open and (b) closed air columns.

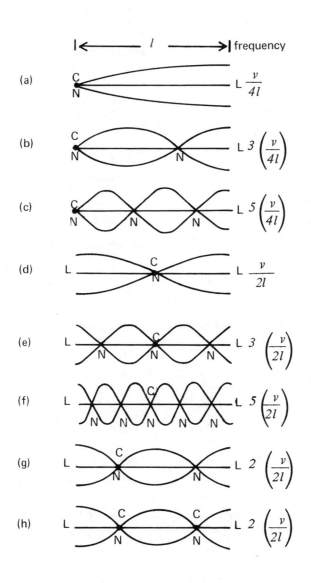

Figure 5.5. Modes of longitudinal vibrations of a rod clamped at positions C.

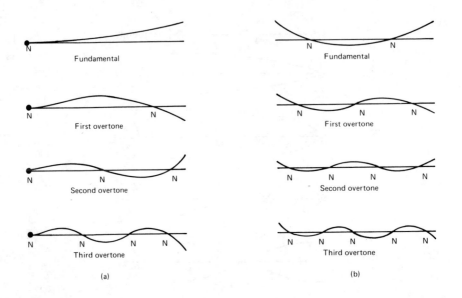

Figure 5.6. Transverse modes of vibration of (a) a bar clamped at one end (b) a bar free at both ends.

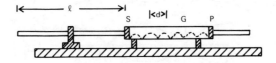

Figure 5.7. Problem 5.14

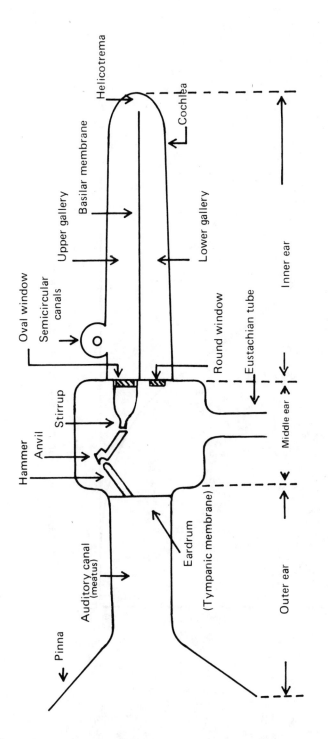

Figure 6.1. Schematic diagram of the human ear. The spiral cochlea is drawn uncoiled for discussion purposes.

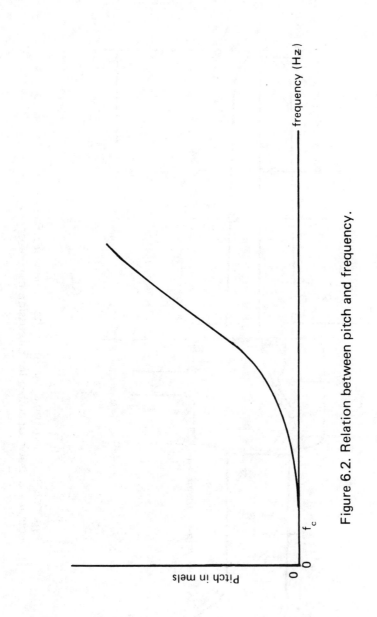

Figure 6.2. Relation between pitch and frequency.

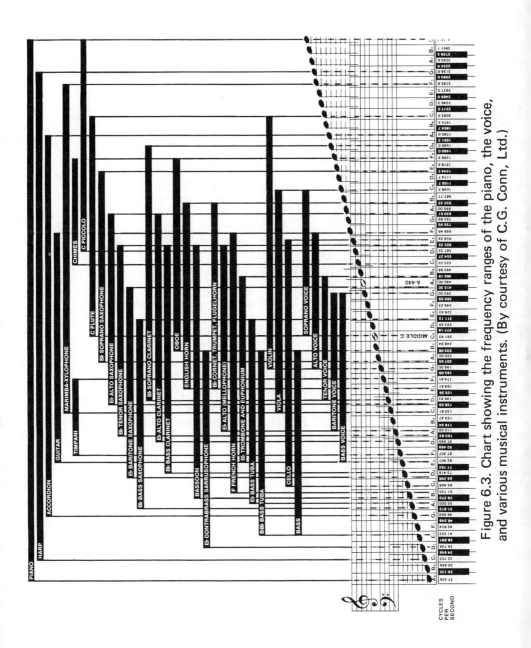

Figure 6.3. Chart showing the frequency ranges of the piano, the voice, and various musical instruments. (By courtesy of C.G. Conn, Ltd.)

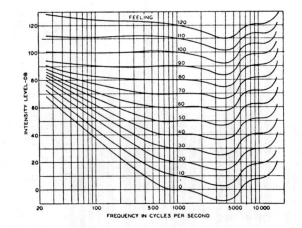

Figure 6.4. Contours of equal loudness
level. The loudness levels are in phons.
(After H. Fletcher and W.A. Munson,
J.A.S.A. 5, 91, 1933. By permission of the
American Institute of Physics.)

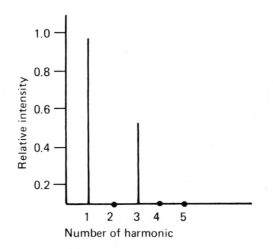

Figure 6.5. Spectrum of a complex wave containing a fundamental and third harmonic whose amplitude is half that of the fundamental.

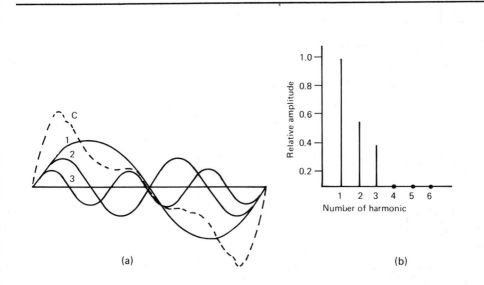

Figure 6.6. (a) Complex wave C composed of a fundamental 1, a second harmonic 2 of half the fundamental's amplitude, and a third harmonic 3 of one-third the fundamental's amplitude. (b) Spectrum of (a).

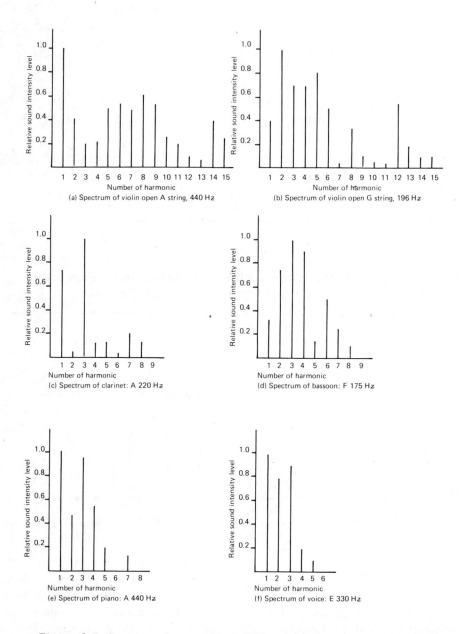

Figure 6.7. Spectra of tones from different musical instruments.

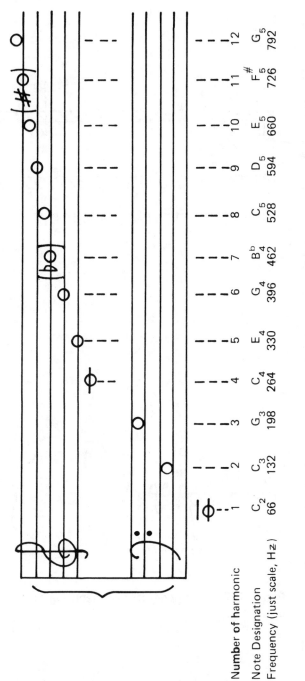

Number of harmonic	1	2	3	4	5	6	7	8	9	10	11	12
Note Designation	C_2	C_3	G_3	C_4	E_4	G_4	B^b_4	C_5	D_5	E_5	$F^{\#}_5$	G_5
Frequency (just scale, Hz)	66	132	198	264	330	396	462	528	594	660	726	792

Figure 7.1. The first twelve partials of the harmonic series based on C_2 as the fundamental.

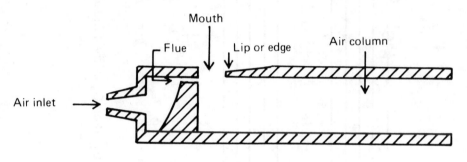

Figure 8.1. Section of a flue organ pipe.

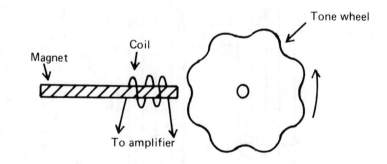

Figure 8.2. Tone generator of electrical organ.

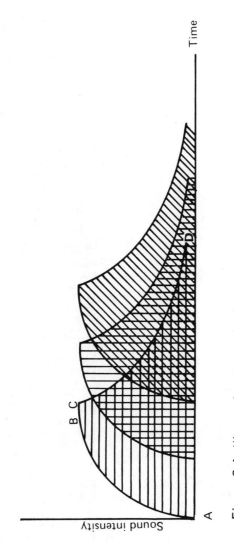
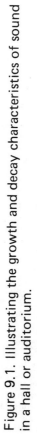

Figure 9.1. Illustrating the growth and decay characteristics of sound in a hall or auditorium.

Index